Alexander Vondrous

Grain growth behavior and efficient large scale simulations of recrystallization with the phase-field method

Schriftenreihe
des Instituts für Angewandte Materialien
Band 44

Karlsruher Institut für Technologie (KIT)
Institut für Angewandte Materialien (IAM)

Eine Übersicht aller bisher in dieser Schriftenreihe erschienenen Bände finden Sie am Ende des Buches.

Grain growth behavior and efficient large scale simulations of recrystallization with the phase-field method

by
Alexander Vondrous

KIT Scientific Publishing

Dissertation, Karlsruher Institut für Technologie (KIT)
Fakultät für Fakultät für Maschinenbau
Tag der mündlichen Prüfung: 9. Juli 2014

Impressum

Scientific Publishing

Karlsruher Institut für Technologie (KIT)
KIT Scientific Publishing
Straße am Forum 2
D-76131 Karlsruhe

KIT Scientific Publishing is a registered trademark of Karlsruhe Institute of Technology. Reprint using the book cover is not allowed.

www.ksp.kit.edu

Print on Demand 2014

ISSN 2192-9963
ISBN 978-3-7315-0280-7
DOI 10.5445/KSP/1000043520

Computational materials science

Grain growth behavior and efficient large scale simulations of recrystallization with the phase-field method

In order to obtain the academic degree
Doctor of Engineering
at the Faculty of Mechanical Engineering
Karlsruhe Institute of Technology (KIT)

Approved dissertation of
M.Sc. Alexander Vondrous

Date of examination: 9.7.2014
Examiner: Prof. Dr. rer. nat. Britta Nestler
Co-examiner: Prof. Dr.-Ing. Dr.-Ing. E.h. A. Erman Tekkaya

Acknowledgments

Prof. Dr. rer. nat. Britta Nestler gave me the opportunity to create this dissertation in her research group during my time as a research assistant. I would like to give her a sincere thanks for supporting my work and for providing the freedom to perform interesting studies.

I also would like to thank my group leader Michael Selzer for discussions and helpful advices. Another warm thank-you goes to my colleagues Mathias Reichardt and Marco Berghoff for creative, idea–rich and constructive cooperation. I also enjoyed working with Johannes Hötzer, Daniel Schneider, Marcus Jainta, Dr.-Ing. Abhik Chouldhury and Dr. Rajdip Mukherjee in a creative and constructive way.

For always patient and helpful support, I thank Claudia Hertweck-Maurer of the secretariat and the system administrators Jonathan Buch and Christof Ratz.

I would like to acknowledge the support provided by my family during the last years, especially the support of my wife Dorota Vondrous.

Abstract

This manuscript summarizes the found insights of grain growth behavior, of multidimensional decomposition for regular grids to efficiently parallelize computing and how to simulate recrystallization by coupling the finite element method with the phase-field method for texture analysis. The frame of the thesis is created by the phase-field method, which is the tool used in this work, to investigate microstructure phenomena.

In the first third of the thesis, the behavior of grain structures during grain growth with random crystal lattice orientations is shown. Isotropic, Read–Shockley and faceted grain boundary energies are compared for two-dimensional simulations of regular n-sided grains with random orientations for $4 \leq n \leq 8$. The analysis shows two properties. In average, the chosen anisotropy models follow the isotropic Neumann–Mullins relation and the growth rate variance reduces with increasing neighbor count. Noticeable fast growth or shrinkage is observed in 1% of the simulations with the faceted grain boundary energy model. A study of the widely used Read–Shockley grain boundary energy model with high resolution (over 268 billion configurations) is performed with a simplified vertex model to obtain the growth rate distributions of regular n-sided grains with $3 \leq n \leq 8$ for the first time.

The second part describes how large-scale simulations with finite differences discretization on high performance computing facilities can be performed efficiently. Dendrite growth, grain coarsening and the growth of ternary eutectics are selected as representative cases of microstructure simulations. Load balancing, domain movement and the influence of multidimensional domain decomposition for parallel execution are analyzed. A timing-based performance model and the performance measurements show a significant faster parallel execution of the multidimensional domain decomposition over one-dimensional domain decomposition due to a better volume to surface ratio.

At the end, the third part is a case study of simulating recrystallization with the phase-field method as part of the virtual process chain hot rolling, cold rolling and annealing in sheet metal production. EBSD data of DC04 sheet metal provided by Dr.-Ing. Simone Schreijäg and Dr. Reiner Mönig of the Karlsruhe Institute of Technology is used to parameterize and validate the three-dimensional recrystallization simulation. The starting condition is the result of a rolling simulation with a crystal plasticity finite element model provided by Pierre Bienger and Dr. Dirk Helm of the Fraunhofer IWM in Freiburg, Germany. The growth and the selection of nuclei is based on the estimated stored energy of subgrain-like grains. Two large-scale 3D simulations are performed on the supercomputer HERMIT in Stuttgart, Germany to obtain the dynamics of the recrystallization process and the final orientation distributions.

Zusammenfassung

Die Doktorarbeit fasst die gewonnenen Erkenntnisse über das Verhalten von Körnern in einer Mikrostruktur zusammen. Besondere Aufmerksamkeit wird auf die mehrdimensionale Zerlegung von regulären Gittern für eine effiziente parallele Datenverarbeitung gelegt und eine Methode zur Simulation von Rekristallisation mit dem Phasenfeldansatz wird vorgestellt.

Im ersten Drittel wird das Verhalten von Kornstrukturen während der Kornvergröberung mit zufällig orientierten Kristallgittern beschrieben. Isotrope, Read–Shockley und facettierte Oberflächenenergie werden für zweidimensionale reguläre Körner mit n Seiten für $4 \leq n \leq 8$ gegenübergestellt. Zwei Eigenschaften zeigt die Analyse. Im Mittel folgen die gewählten Modelle für anisotrope Oberflächenenergie der isotropen Neumann–Mullins Relation und die Varianz der Wachstumsrate nimmt mit steigender Anzahl an Nachbarn ab. Auffällig schnelles Kornwachstum wurde bei 1% der Simulationen mit der facettierten Oberflächenenergie beobachtet. Das weit verbreitete und gerne verwendet Read–Shockley Modell für Korngrenzenergie wird mit einem vereinfachten Vertexmodell, hoher Auflösung (über 268 Milliarden Konfigurationen) und für reguläre Körner mit n-Seiten mit $3 \leq n \leq 8$ untersucht, um die Verteilung der Wachstumsraten zum ersten Mal darzustellen.

Das zweite Drittel beschreibt, wie große Simulationen mit der finiten Differenzen Methode auf Hochleistungsrechnern effizient durchgeführt werden können. Dendritenwachstum, Kornvergröberung und das Wachstum von ternären Eutektika dienen als repräsentative Beispiele für Mikrostruktursimulationen zur Analyse verschiedener Optimierungstechniken. Lastverteilung, ein sich bewegendes Simulationsgebiet und der Einfluss mehrdimensionaler Gebietszerlegung wird analysiert. Das Performance-Modell und die Messergebnisse zeigen eine schnellere parallele Verarbeitung bei mehrdimensionaler Zerlegung gegenüber eindimensionaler Zerlegung.

Den Abschluss bildet eine Fallstudie zur Simulation von Rekristallisation mit der Phasenfeldmethode als Teil der Prozesskette Warmwalzen, Kaltwalzen und Glühen in der Blechproduktion. EBSD-Daten eines DC04 Bleches wurden von Dr.-Ing. Simone Schreijäg und Dr. Reiner Mönig des Karlsruher Instituts für Technologie für die Parametrisierung und Validierung der dreidimensionalen Reskristallisationssimulaiton genutzt. Die Startbedingung ist das Ergebnis einer Kaltwalzsimulation mit der Finite Elemente Methode auf Basis eines Kristallplastizitätsmodells von Pierre Bienger und Dr. Dirk Helm vom Fraunhofer IWM in Freiburg. Für die Auswahl der Keime und für das Wachstum der Keime wird die gespeicherte Energie und die Subkornstruktur genutzt. Eine große dreidimensionale Simulation wurde auf dem Supercomputer HERMIT in Stuttgart durchgeführt, um die Dynamik der Mikrostruktur bei der Rekristallisation zu erhalten.

Contents

1. Introduction

A short historical introduction illustrates the long history of over 10 000 years of metal related materials science and the importance of materials science for our civilization until today. Modern civilization is characterized by the use of computers, which changed the way scientific questions are answered and asked. This influence of computers on scientific work is introduced and related to the topics of this thesis.

The knowledge about material properties, how to process materials and to apply this knowledge to a wide variety of applications, is closely related to the history of mankind. Wood, stone, fibers and ceramics are the first materials humans have used because of their natural occurrence and accessibility. Little is known about how the first metal smelts have been invented to extract metal from ore. Ceramic manufacturing, which dates back to about 28 000 BC [43] seems to be related to the discovery of smelts and metal extraction because of burning clay vessels with fire.

The use of metals can be dated back to about 8 000 BC. Natural sources of metals are rare and often in the form of meteorites. The first smelts to extract copper from ore, like malachite and azurite, date back to 5 000 BC (copper age) in today's Turkey. An understanding of the mechanical properties like toughness, ductility and elasticity grew with the knowledge of metal production and processing during the next 5 000 years from copper over bronze to iron.

Iron was a by-product during copper production in the form of sponge iron with less favorable mechanical properties even in the pure form compared to bronze. The great Hettites empire (1400 to 1200 BC) span from central Turkey over northern Syria to northern Iraq. They were the first who refined iron production and carburized iron to increase the toughness. The collapse of the Hettites empire in around 1160 BC marked the beginning of the iron age at 1200 BC to about 600 BC. Iron replaced bronze as the favorable metal for tools and weapons due to better mechanical properties and a higher resource security.

Until today, refinement of and investigations in production and processing of materials like steel or aluminum has been crucial for the health and wealth of modern civilizations. The following example shows the necessity for a better understanding of material properties to improve living conditions world wide. Figure 1.1 relates the world crude steel production[1] to the world population[2] from 1900 to 2012, respectively, 2015. Without discussing the details, figure 1.1 shows that a bigger population has a higher resource demand and population growth is not expected to stop soon.

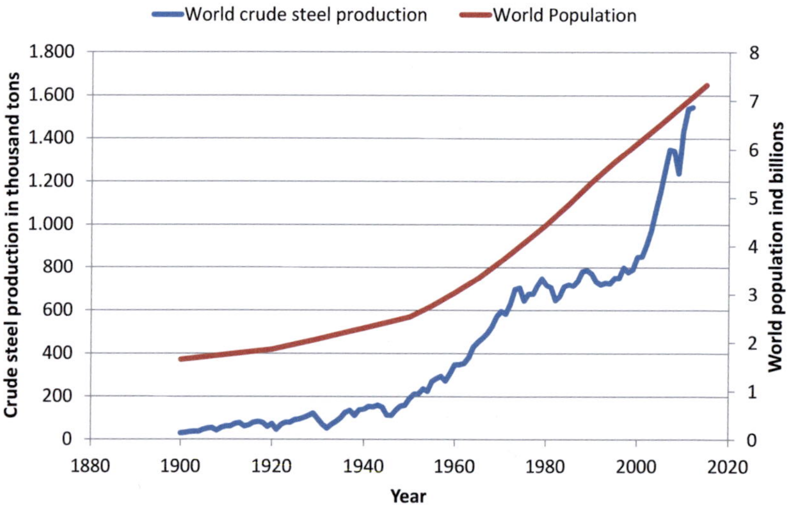

Figure 1.1.: World crude steel production and world population from 1900 until today shows the relation, that more humans need more steel and more resources, respectively.

Health, wealth, knowledge gain (science) and the dissemination of knowledge (education) are necessary to tackle difficulties of our civilization now

[1]World Steel Association Statistical Yearbooks from 1978 to 2013

[2]United Nations - Population Division, Department of Economic and Social Affairs

and tomorrow. One small step towards a better world is to gain knowledge about methods to simulate material properties efficiently.

The theory that all matter has to be constructed from small indivisible parts is called atomism and dates back to the ancient Greeks around 600 BC. Philosophers questioned the structure of matter and they gave a well thought out answer without the technical equipment of today. In the 18th and 19th century scientists adopted the Greek word ἄτομος (atomos), which means indivisible, to name the smallest parts they could produce. Unfortunately, atoms and their components consist of even smaller particles, which was discovered in the 20th century. During that time scientists developed theories and experiments to explain and proof the physical properties of matter, time and space.

Another development beside the rise of modern physics in the 20th century changed the way scientific questions are answered and asked. Computer simulations became the third pillar of science next to experiment and theory [29, 18], because they make observations accessible, if experimental observations are not possible because of extreme scales, health hazards and technical or ethical reasons. This new tool also introduced new scientific fields such as artificial intelligence, cryptography or fields like computational materials science. The latter scientific field is the area wherein this thesis is settled.

Computational materials science is computer centered materials science and includes a wide range of topics like algorithms to e. g. generate, analyze and characterize structures or methods and models to simulate material behavior. The interdisciplinary character can make computational materials science difficult. Simulations are a good example to explain the inherent difficulty, because simulations combine the three disciplines physics, maths and informatics. To overcome this obstacle, interdisciplinary teams are formed to develop state-of-the-art models and implementations to reproduce physics efficiently with computers.

As indicated by the brief historical review in this section, the field of study are metals, in particular the simulation of grains and grain structures at the mesoscopic scale with the phase-field method. A short introduction into grain structures, the phase-field method, the Voronoi tessellation and parallel computing is given to introduce and motivate the main concepts and tools.

1.1. Metallic grain structures

Metals consist of atoms, which are regularly structured in one of seven crystal systems (32 crystal classes). A body-centered cubic crystal structure with red atoms is depicted in figure 1.2a as an example. Most of the products built from metals do not consist of a single crystal as e. g. turbine blades, because the production process is resource intense and the thermomechanical requirements are not as high as for turbine blades. A polycrystalline structure is mostly present in metal products, which consists of groups of atoms with the same crystal lattice but with different orientations. The group of atoms in the same crystal lattice is referred to as a grain. A generated grain structure in two dimensions is depicted in figure 1.2b where blue cubes indicate the crystal lattice orientation of each grain.

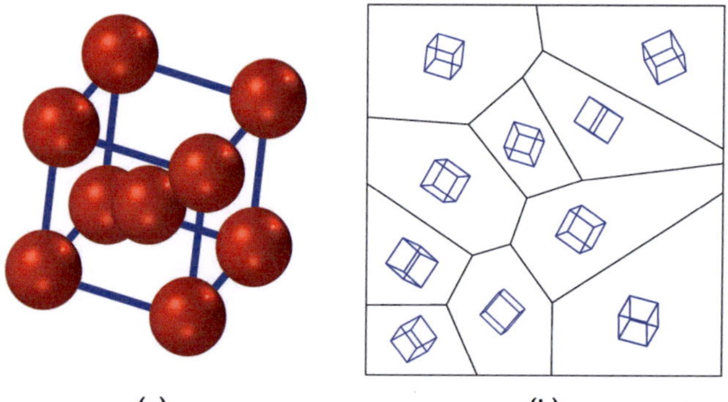

(a) (b)

Figure 1.2.: a) Body-centered cubic crystal lattice with red atoms. b) A grain structure (black lines) with differently orientated cubic crystal lattices indicated by blue cubes.

The interface between two grains (grain boundary) consists of atoms, which do not reside on one of the adjacent crystal lattices, such that this region has special properties. A faster diffusion process than in the bulk region or defect pile-up at grain boundaries during deformation are two examples. The defect pile-up mechanism is used to increase the strength of alloys as in oxide dispersion-strengthened alloys, which have a fine grain structure and hence a big grain boundary area. Because the atoms of grain boundaries are in an energetically higher state, a grain structure is trying to remove the grain boundary area or is trying to reach an equilibrium state of the grain boundary energies. The energy of a grain boundary is also called surface tension, which is dependent on the orientations of the crystal lattices and the orientation of the grain boundary. A description of the grain boundaries and their behavior for isotropic and anisotropic grain boundary energies is presented in the chapters 3 and 4.

During the production of e. g. sheet metal, it is necessary to control the orientation distribution of the crystal lattices beside the amount of crystal lattice defects and the amount of grain boundary area to define the mechanical properties. As one can imagine, the mechanical response on a cubic crystal is dependent on the orientations of the lattice and the force, such that a grain structure with crystal lattices oriented in a preferred direction will behave anisotropically on mechanical load. A preferred orientation distribution can occur during production as e. g. during cold rolling of sheet metal. Certain crystal lattices are rotated more easily into a specific direction than others during rolling, such that some orientations have a higher probability. One mechanism to modify the anisotropy is by recrystallization, where a new grain structure appears during annealing of a plastically deformed metal piece. A description of recrystallization and the preferred orientations after cold rolling and recrystallization is presented in chapter 7.

The presence of preferred orientations can be visualized either by pole figures or plots of the orientation distribution function in Euler space. Both ways of visualization present a heat map-like figure with a hot region, where the orientations accumulate. The Euler space representation of the orientation distribution function is used in this thesis to describe the influence of cold rolling and subsequent annealing on the orientation distribution.

Another important mechanism to manipulate the properties of metal products is by alloying and heat treatment. The iron carbon composition is the most common example to illustrate the influence of a second element and temperature on a metal structure. At low temperatures beneath 600°C and 6% carbon, a body-centered cubic crystal structure is present. At higher temperatures, the crystal lattice transforms to a face-centered cubic lattice or to cementite, which is an orthorhombic lattice of Fe_3C. The relation of heat and composition is presented in phase diagrams, which exist for various alloys and elements. In this context, a phase represents a defined crystallographic configuration or defined mixture of configurations. In this thesis, single phase grain structures with constant temperature are assumed to study grain growth and recrystallization at a large scale.

1.2. The phase-field method

In contrast to the finite element method or the finite difference method, the phase-field method is not a numerical discretization method, it is rather a method to model free boundaries and bulk properties. It is often used to simulate microstructure dynamics under various conditions, as reviewed by Chen in [12].

The phase-field method is characterized by a vector-valued order parameter to describe a diffuse interface, as depicted in figure 1.3. A phase is one component of the vector-valued order parameter and describes a physical state. Its name is misleading because it must not be one phase of a phase diagram. One phase represents for example one grain with a specific lattice orientation, such that each grain of a grain structure is represented by its own phase. Unfortunately, the simulation of a large grain structure with over 200 000 grains is not possible without a dynamic phase vector, as presented by Patnaik and Vedantam in [116], by Kim et al. in [52] or by Nestler et al. in [81].

Originally, the phase-field method used only two phases, where the second phase is computed by one minus the first phase due to the sum constraint of the order parameter. Now, multiphase and multicomponent models are available to describe complex systems, as presented e. g. by Nestler and Garcke in [80]. In addition, one advantage of the phase-field method is that an explicit tracking of the interface is not necessary as for vertex models or most finite element-based models; the smooth coupling with other models like fluid dynamics, magnetism or mechanical fields is also a specific feature of the phase-field method.

The phase-field method is formulated by a functional, which either describes energy or entropy densities. The functional for this thesis is the energy density functional without a contribution of a bulk energy density, such that only interface energy densities are considered, because grain growth and recrystallization are surface driven processes. Chapter 3 describes the phase-field model in detail and chapter 7 explains an additional energy density to simulate recrystallization.

The PACE3D (Parallel Algorithms for Crystal Evolution in 3D) framework contains a collection of tools and model implementation is used as a code base, which implements the phase-field model of Nestler and Garcke. PACE3D contains tools for pre- and post-processing simulations as well as many models to simulate complex systems. The core of PACE3D is the multiphase and multicomponent model of Nestler and Garcke [80], which is implemented modularly to preserve flexibility and the ease of use. All features, model extensions and tools presented or used in this thesis are implemented in the framework of PACE3D, which is mainly written in the programming language C.

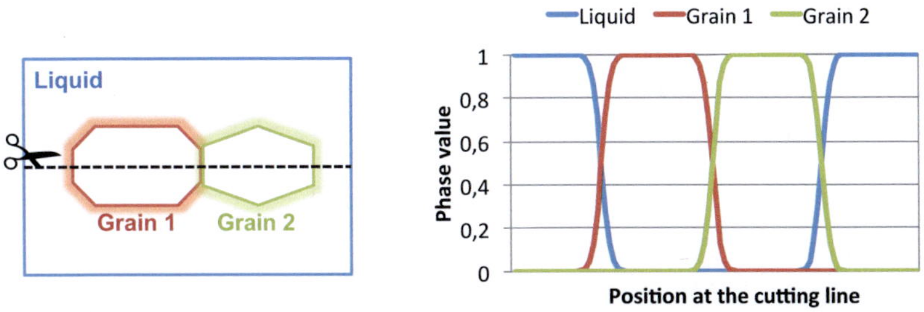

Figure 1.3.: On the left side, there is a two-dimensional liquid melt with two solid grains. The phase values along the dashed line are plotted in the diagram on the right side for each phase separately.

1.3. Voronoi tessellation

Because the Voronoi tessellation is used often in this work, a short introduction is given to present the concept and the main properties. The Voronoi tessellation is a common and flexible tessellation method to generate grain structure-like starting conditions. In order to simulate grain structures with specific topological properties, it is useful to be able to generate starting conditions, if no, less or inappropriate experimental measurements or other restrictions are present.

The basic idea is to divide a domain into subdomains, called Voronoi cells, by Voronoi points, which are often placed randomly. A Voronoi point consists of the spatial coordinates, a number and sometimes a weight to define the Voronoi cell. A Voronoi cell divides the spatial domain in subdomains where each subdomain encloses the area of the domain which is nearest to a Voronoi point. Figure 1.4 depicts a two-dimensional example with 20 Voronoi points in blue and their Voronoi cells, whose boundaries are black lines.

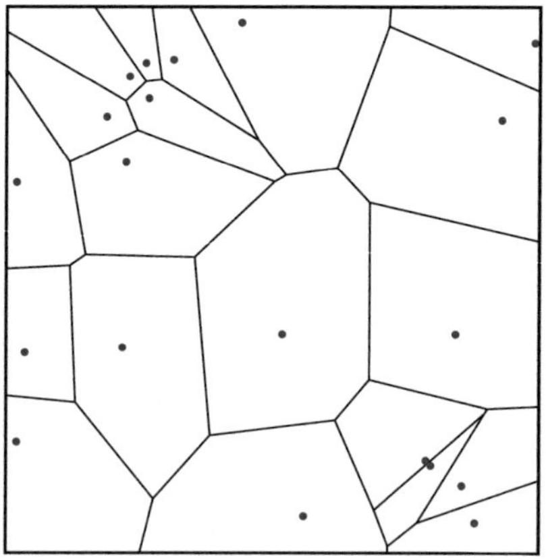

Figure 1.4.: A Voronoi tessellated, two-dimensional domain with blue Voronoi points. A cluster at the bottom right corner generates one small Voronoi cell with a sharp angle, which is not common in grain structures.

The following short algorithmic description of the Voronoi tessellation, based on a regular grid, explains the Voronoi tessellation from another perspective.

```
generate random voronoi points;
for each lattice point do
    get nearest voronoi point to current lattice point;
    assign number of nearest voronoi point to lattice point;
end
```

Algorithm 1: Basic Voronoi algorithm on regular grids.

Some minor modifications of the basic algorithm enable the generation of many very different structures [5]. With a specific placement of Voronoi points, for example, all convex structures can be generated by placing the Voronoi points at the vertices of the dual form of the convex structure. This mechanism is used in the chapters 3 and 4 to create regular n-sided grains.

The following examples show the flexibility of the Voronoi algorithm to generate grain structures with specific properties. An iterative method enables the generation of globular Voronoi cells by placing the random Voronoi points after tessellation in their Voronoi cell barycenters, as described in algorithm 2.

```
generate random voronoi points;
for n iterations do
    for each lattice point do
        get nearest voronoi point to current lattice point;
        assign number of nearest voronoi point to lattice point;
    end
    compute barry center of each voronoi cell;
    place each voronoi point at its voronoi cell barycenter;
end
```

Algorithm 2: Voronoi algorithm to generate globular Voronoi cells by placing the Voronoi points in their Voronoi cell barycenter.

In order to generate a specific size distribution, as presented by Balzer and Deussen in [8], the iterative algorithm needs to be extended by assigning each Voronoi point a weight. This weight is used to manipulate the distance metric, such that large Voronoi cells have a high weight and small Voronoi cells have a small weight. A large collection of metrics is available, such that curved or faceted Voronoi cells can be generated.

The basic Voronoi tessellation is used in this work to generate regular n-sided grains in two dimensions and to map three-dimensional finite element data onto a regular grid.

1.4. Parallel computing

The fast pace of the semiconductor development since around 1960 still keeps its momentum of almost doubling the amount of integrated circuits per area about every two years [76]. Until around 2003, ordinary PCs and workstations gained performance by an increasing clock frequency and the technical development of the CPU architecture. Around 2003, the clock frequencies stagnated because it is difficult to integrate e. g. liquid nitrogen chillers on top of processors economically. Processors with two or four CPUs at lower frequencies allowed to increase the computation performance without cooling problems. Moore's law (doubling the integrated circuits per area almost every two years) is expected to be true for the next decade, such that even more CPUs per computer can be expected.

This development is not only of great interest for scientific fields, which rely on computation power or where computation power is always lacking. All scientific applications have to be able to use several CPUs concurrently in order to benefit from new generations of processors. Another good reason to care about and to consider parallel computing is the opportunity to switch to a supercomputer if the requirements increase. A supercomputer is not only able to compute a large task fast, it is also able to provide the necessary memory for memory intense applications.

In order to estimate the computational power of workstations in five or ten years, it is helpful to understand the history. Figure 1.5 shows the performance of the fastest and the slowest supercomputer of the TOP500 list[3] from 1993 until today. The performance is measured in floating point

[3] The 500 fastest computers of the world since 1993: http://www.top500.org

operations per second (flop/s) with metric prefixes. The fastest computers drop out of the list after about eight years and it takes about ten or eleven years to have the computation power of the slowest supercomputer available at home. Currently, the slowest supercomputer has 22 212 CPUs and 117.8 Tflop/s.

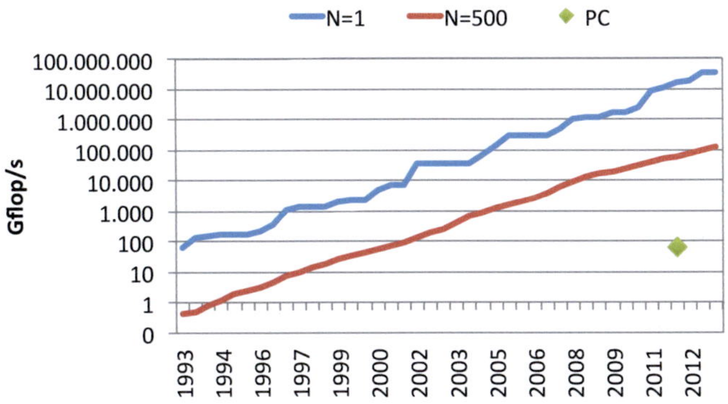

Figure 1.5.: The performance of the fastest and slowest supercomputer in the top 500 list compared to the performance of a PC in 2012.

Parallelization of an application involves the decomposition of a computation task, such that each CPU is able to compute its task independently. Dependencies are resolved by communication, which reduces the parallel efficiency if not managed correctly. Communication is less a bottleneck for parallel computations on a computer where a shared memory can be used to communicate between CPUs. On supercomputers or high performance clusters with high performance networks, it is necessary to minimize the amount of data and the distance because network technology does not keep pace with the processor development.

Many parallelization libraries and concepts are available. Almost every modern programming language supports threading, which allows to parallelize an application on a single computer (shared memory parallelization). For simplicity, OpenMP was invented to hide thread management for the programming languages C and C++. To utilize a supercomputer, which consists of many connected high performance computers (distributed memory parallelization), it is common to send messages with one implementation of the message passing interface (MPI) standard [73]. The minority of supercomputers consist mainly of general purpose graphics processing units (GPGPU) to make use of the large and vectorized floating point units. Those supercomputers are programmed either with OpenCL or CUDA.

Parallelization in PACE3D is performed with MPI by decomposing a regular grid into subdomains, which are computed independently. Chapter 5 and 6 describe and discuss the parallelization scheme in detail.

2. Research Questions

This work is not referring to solely one research question, rather to one topic: Grain structure simulations with the phase-field method. The research questions are related to modeling, efficient parallel execution and to applying the phase-field method to simulate recrystallization. In particular, the following questions are answered or worked on:

- How do the Read–Shockley and faceted grain boundary energy models influence grain growth?
- What do the growth rate distributions with the Read–Shockley grain boundary energy model for 3- to 8-sided regular grains in two dimensions look like?
- How to perform efficient large-scale 3D phase-field simulations on regular finite difference grids with a supercomputer?
- How to estimate the parallel runtime of a finite differences discretization scheme with dimensional domain decomposition?
- How to model and simulate static primary recrystallization of a 3D microstructure with the phase-field method as part of a numerical process chain in sheet metal production?

3. Grain Growth Behavior

A computational study with over 1 000 two-dimensional (2D) phase-field simulations is performed to show the isotropic character of the Read–Shockley and faceted surface energy anisotropy models for random crystal lattices. The two anisotropic surface energy models follow the analytical Neumann–Mullins growth law for isotropic surface energies in average with random crystal lattices. Another common property is the reduced standard deviation of the growth rates for an increasing neighbor count.

The properties of grain boundary energy models must be well known to model the behavior of grains and grain structures without surprising side effects. Extreme cases of grain growth or shrinkage must be known to e. g. avoid abnormally growing grains in cases where abnormal growth cannot occur. A good example is the sintering of components for the metal processing industry, to produce the cutting edges of machining tools with tungsten and titanium carbid, which requires an isotropic and a faceted grain boundary energy model for a proper simulation [83, 107, 131]. Figure 3.1 shows round isotropic and strongly faceted grains of a TiC-WC-Co hard metal microstructure.

The simplest model for grain boundary energy is an isotropic formulation. The energy for all grain boundaries is the same. Connected soap bubbles and froth are examples with isotropic boundary energy. Analytical models to determine the growth rate of a bubble surrounded by n-neighboring bubbles exist for two-dimensional (2D) and three-dimensional (3D) structures. The Neumann–Mullins relation was formulated in 1952 [78] and explains the relation between the number of triple junctions and the growth rate for 2D grains or structures with isotropic surface energy. Half a century later, in 2007, a formulation for 3D structures was formulated by McPherson and Srolovitz [69], which describes the relation between the growth rate, triple line length and mean width. These analytical models are suitable to determine the growth rates of structures with isotropic or averaged surface energy. Another part of structures has stronger anisotropic properties, like in crystal shape engineering [66]. Averaging results in strong deviations from the isotropic solutions, such that the Neumann–Mullins or McPherson–Srolovitz models cannot be applied to investigate anisotropy effects.

Figure 3.1.: Sintered TiC-WC-Co microstructure with isotropic and faceted grains. An abnormally grown faceted grain is located in the center. This picture is a copy of figure 2 f) from [131] and is kindly provided by Suk-Joong L. Kang.

Anisotropic grain boundary energies have too many degrees of freedom, to easily find a general analytical solution. As mentioned by Rohrer in [96], the grain boundary can be described by five parameters, where three are used to relate two crystal lattices to each other, and two parameters to describe the boundary plane. Depending on the anisotropy model and due to the huge parameter space in anisotropic materials, a broad variety of effects is observed. Rohrer [96] reviewed the important role of anisotropy models to design and to understand grain structure properties, despite the high complexity of anisotropy modeling, which motivates the investigation considered in this and the following chapter.

The semi-anisotropic Read–Shockley model [93] is experimentally validated for tilt grain boundaries of cubic crystal lattices and angles smaller than 30°. The model is based on dislocations accumulating in the boundary until the complexity of the dislocations interactions is too high. As an experimentally determined function with one parameter (misorientation), it is attractive for modelling purposes, as e. g. performed in [32]. The only parameter is the crystal lattice misorientation angle between the grains at the grain boundary. The grain boundary direction and the full crystal lattice relation is not considered.

Another well-known grain behavior is faceted growth of crystals. George Wulff showed in 1901 [129] how to determine the crystal shape for a given surface free energy. The solution is called Wulff construction and it determines the crystal shape geometrically with the polar plot of the energy. The nice review of Conyers Herring [40] about the Wulff construction explains the ability to model round sharp edges for crystals depending on the application. Faceted growth models for the phase-field method exist and are described in [114, 19, 115, 68]. These models are used to investigate thin film growth of zeolites [125], dendrite growth [30] or directional solidification [64].

Each of the three briefly introduced grain growth phenomena has its own characteristic influence on the microstructure. The idea behind this study is to obtain properties like maximum, average and minimum growth rates by scanning a small part of the parameter space to characterize growth behavior of the Read–Shockley and the faceted anisotropy models. Five configurations of regular n-sided grains in 2D with $4 \leq n \leq 8$ and 100 different lattice arrangements are used. Figure 3.2 shows the five topology classes with the blue colored grains of interest. Detailed descriptions of the isotropic, Read–Shockley and faceted grain growth are given in chapter 3.2.1, 3.2.2 and 3.2.3. The simulations and development of additional tools is performed in the framework provided by the PACE3D (Parallel Algorithms for Crystal Evolution in 3D) collection of tools.

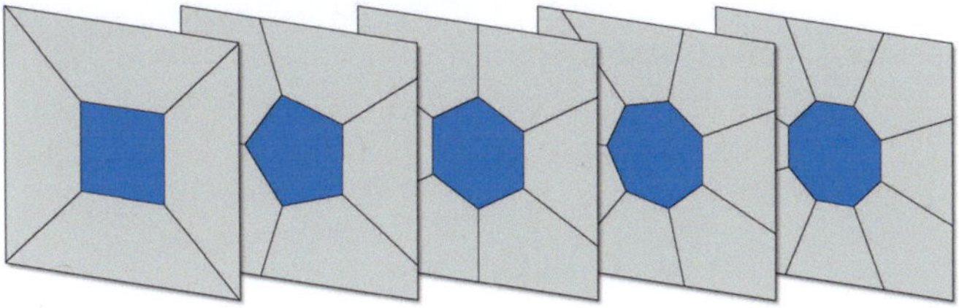

Figure 3.2.: All starting conditions for the phase-field simulations from 4 to 8 neighbors.

3.1. Phase-Field Modeling

The original model of Nestler, Garcke and Stinner [80] is created to simulate multiphase and multicomponent systems. Grain coarsening is surface driven, such that an evolution equation for the phase-field vector $\boldsymbol{\phi}$ is derived, describing each grain state in the simulation domain Ω from an energy functional of the form:

$$\mathcal{F}(\boldsymbol{\phi}) = \int_{\Omega} \varepsilon a(\boldsymbol{\phi}, \nabla\boldsymbol{\phi}) + \frac{1}{\varepsilon} w(\boldsymbol{\phi}) dx. \tag{3.1}$$

The functional consists of the surface energy density function $a(\boldsymbol{\phi}, \nabla\boldsymbol{\phi})$ and of the multi-obstacle potential $w(\boldsymbol{\phi})$. Each phase represents one grain in the system of N grains. The surface energy density function describes the interface energy by a summation of pairwise contributions of grains i and j and reads as

$$a(\boldsymbol{\phi}, \nabla\boldsymbol{\phi}) = \sum_{i<j} \gamma_{ij} [a_{ij}(q_{ij})]^2 |q_{ij}|. \tag{3.2}$$

In this expression, the generalized gradient vector is

$$q_{ij} = \phi_i \nabla\phi_j - \phi_j \nabla\phi_i. \tag{3.3}$$

The grain boundary energy between the grains i and j ($i \neq j$ and $1 \leq i, j \leq N$) is γ_{ij} and the grain boundary mobility between the two grains i and j is denoted by τ_{ij}^{-1}. A description of the anisotropy model properties using γ_{ij} and τ_{ij} is given in the chapters 3.2.1, 3.2.2 and 3.2.3. For the isotropic and semi-anisotropic (Read–Shockley) grain boundary energy model $a_{ij}(q_{ij})$ is set to 1, as described in [80]. For strongly faceted anisotropy, η_{ij}^k are the corner vectors of the crystal shape with n_{ij} corners and $a_{ij}(q_{ij})$ is

$$a_{ij}(q_{ij}) = \max_{1 \leq k \leq n_{ij}} \left\{ \frac{q_{ij}}{|q_{ij}|} \eta_{ij}^k \right\}. \tag{3.4}$$

The multi-obstacle potential has the form

$$w(\boldsymbol{\phi}) = \frac{16}{\pi^2} \sum_{i<j} \gamma_{ij} \phi_i \phi_j + \sum_{i<j<\delta} \gamma_{ij\delta} \phi_i \phi_j \phi_\delta \tag{3.5}$$

and contains a higher order term, to suppress the occurrence of artificial grain contributions at two grain boundaries. The surface energy function and the multi-obstacle potential are scaled with ε, which determines the diffuse interface width. ε is selected in such a way, that 10 to 15 support points are in the interface region. To derive the evolution equations for ϕ_i, the variational derivatives are derived in the form

$$\tau\epsilon\frac{\partial\phi_i}{\partial t} = -\frac{\delta\mathcal{F}}{\delta\phi_i} - \lambda, \tag{3.6}$$

$$\tau\varepsilon\frac{\partial\phi_i}{\partial t} = \varepsilon[\nabla\cdot a_{,\nabla\phi_i}(\boldsymbol{\phi},\nabla\boldsymbol{\phi}) - a_{,\phi_i}(\boldsymbol{\phi},\nabla\boldsymbol{\phi})] - \frac{1}{\varepsilon}w_{,\phi_i}(\boldsymbol{\phi}) - \lambda. \tag{3.7}$$

Further, the notation $a_{,\phi_i}$, $a_{,\nabla\phi_i}$ and $w_{,\phi_i}$ for the derivatives of the energy contributions, with respect to ϕ_i and $\nabla\phi_i$, is used. For all locally occurring grains $l \leq N$, the constraint $\sum_{i=1}^{N}\phi_i = 1$ is guaranteed by the Lagrange multiplier λ:

$$\lambda = \frac{1}{l}\sum_{i=1}^{l}\left[\varepsilon[\nabla\cdot a_{,\nabla\phi_i}(\boldsymbol{\phi},\nabla\boldsymbol{\phi}) - a_{,\phi_i}(\boldsymbol{\phi},\nabla\boldsymbol{\phi})] - \frac{1}{\varepsilon}w_{,\phi_i}(\boldsymbol{\phi})\right] \tag{3.8}$$

In equation (3.7), τ is the averaged value for all local mobilities τ_{ij} and is defined by:

$$\tau = \frac{\sum_{i<j}\phi_i\phi_j\tau_{ij}}{\sum_{i<j}\phi_i\phi_j} \tag{3.9}$$

The averaging is important to determine the mobility of triple junctions in 2D or triple lines and quadruple junctions in 3D.

3.2. Grain boundary energy formulations

A notation for 2D grains is introduced to explain the relations between grain boundary energy, triple junctions and grain volume for the following sections. The grains Φ_1, Φ_2 and Φ_3 have common grain boundaries, GB_{12}, GB_{13} and GB_{23}, which meet at the triple junction TJ_{123}. Figure 3.3 shows a regular 6-sided grain Φ^6 with 6 adjacent triple junctions.

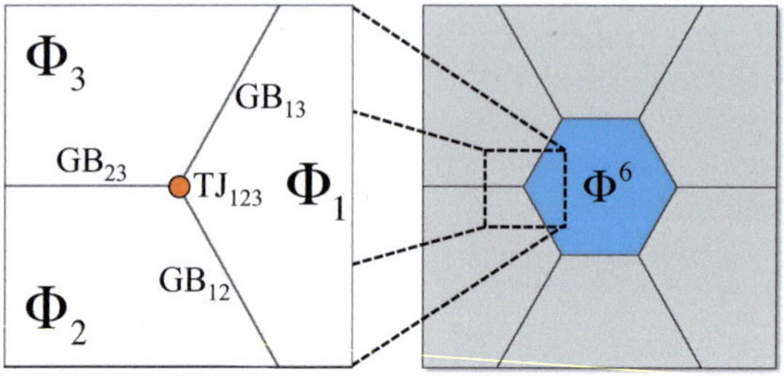

Figure 3.3.: The orange circle is a triple junction with adjacent grain boundaries cut out of the 6-sided grain.

It must be noted, that equation (3.4) is set to 1 for the isotropic and the semi-anisotropic (Read–Shockley) grain boundary energy model. The faceted anisotropy uses a γ_{ij} of 1 for all grain boundaries.

3.2.1. Isotropic grain boundary energy

A system is called isotropic if all grain boundaries have the same grain boundary energy. The Herring equation [39] is a general description of the equilibrium state of the grain boundary where the sum of all forces tangential and normal to the grain boundary has to be zero, as outlined in figure 3.4. If all grain boundary energies are equal and normal forces are not considered, the simplified Herring equations, also known as Young's equations, can be used to determine the equilibrium angles between the grain boundaries at a triple junction. In 2D the equilibrium angle between the grain boundaries at a triple junction is 120°. In 3D the equilibrium angle between triple lines at a quadruple point is 109.5°.

If the grain boundaries at a triple junction do not have 120°, the triple junctions move to reach equilibrium. A grain in 2D with six neighbors, as shown in figure 3.3, is in equilibrium, because all angles are 120°. If a grain has less than six neighbors, the angle between GB_{12} and GB_{13} (see figure 3.3) is less than 120°. Without considering normal forces, the tangential forces pull the triple junction towards the grain center. For more than six neighbors the grain will grow, because the tangential forces pull the triple junction away from the grain center. This relation is described by the Neumann–Mullins equation [78].

In the phase-field framework, the grain boundary energy γ_{ij} and the grain boundary mobility τ_{ij}^{-1} are set to constant values for all phase combinations (grain boundaries) to simulate isotropic grain growth.

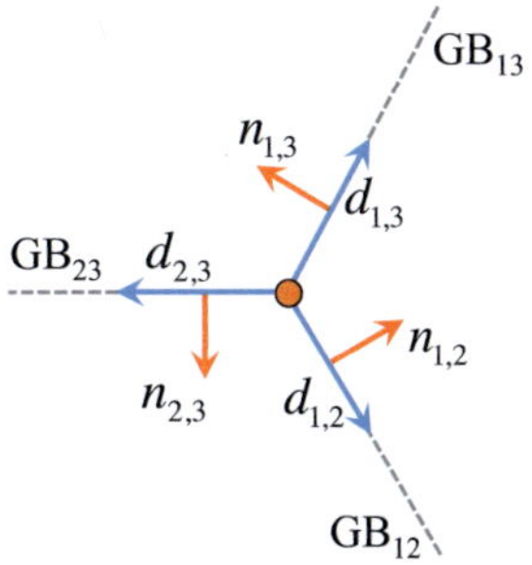

Figure 3.4.: All force vectors of the Herring equation tangential $d_{i,j}$ (blue) and normal $n_{i,j}$ (orange) to the dashed grain boundaries (grey).

3.2.2. Read–Shockley grain boundary energy

One step away from a purely isotropic system is to define a specific energy for each grain boundary. The next question is, which grain boundary energies are present? Read and Shockley derived a dislocation-based model for [100] tilt angle grain boundaries smaller than 30° and materials with cubic crystal lattice. This restricted Read–Shockley model is applied for simulation purposes by expanding it to all kinds of grain boundaries like [110] tilt or twist grain boundaries, which is a significant simplification. One further assumption is, that above a high angle grain boundary threshold Θ_{max}, the energy remains constant. The misorientation angle Θ_{ij} between two crystal lattices is used as the input parameter to compute the energy between two grains, such that a common[4] form of the Read–Shockleys grain boundary energy is

$$\gamma_{ij}(\Theta_{ij}) = \begin{cases} \frac{\Theta_{ij}}{\Theta_{\text{max}}} \cdot (1 - \ln(\frac{\Theta_{ij}}{\Theta_{\text{max}}})) & \text{if } \Theta_{ij} < \Theta_{\text{max}} \\ 1 & \text{otherwise} \end{cases} \tag{3.10}$$

Grain boundary mobility is used in the same manner. High angle grain boundaries above 15° have a higher mobility. In contrast to the grain boundary energy, the mobility is modeled with a sigmoidal function like in [44]. Equation (3.11) is the reciprocal mobility.

$$\tau_{ij}(\Theta_{ij}) = \frac{1}{1 + \frac{0.9}{1+\exp(-2(\Theta_{ij}-\Theta_{\text{max}}))}} \tag{3.11}$$

A cubic crystal lattice and rotations in the 2D plane are used, such that the biggest misorientation angle is 45°. Due to a missing third dimension only [100] tilt angle grain boundaries are present as modeled by Read and Shockley. Figure 3.5 is a plot of the grain boundary energy and the mobility over the full misorientation range from 0° to 45° with $\Theta_{\text{max}} = 15°$. One distinct feature of the Read–Shockley model with Θ_{max} of 15° is the large isotropic part with 60%.

The Read–Shockley model is a semi-anisotropic model because the grain boundary orientation q_{ij} is not considered.

[4]The work of Humphreys [44] is cited by about 267 papers according to the Scopus database and the work of Read and Shockley [93] is cited by about 717 papers according to APS physics.

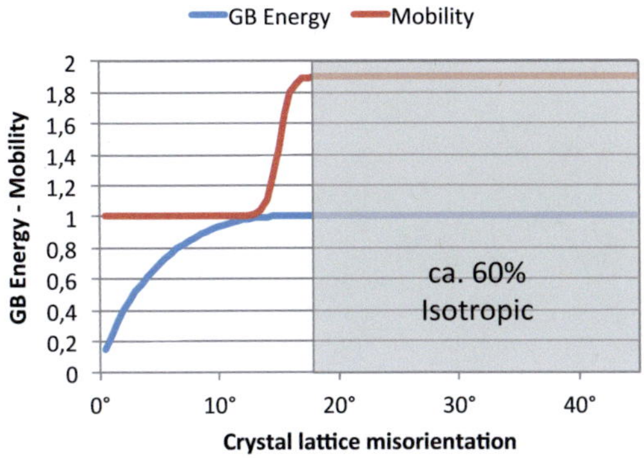

Figure 3.5.: Grain boundary energy and mobility of the Read–Shockley model over the misorientation angle. About 60% of the Read–Shockley model is isotropic for $\Theta_{\max} = 15°$.

3.2.3. Faceted anisotropy

The last anisotropy model is based on the surface free energy (Gibbs free energy). To identify the surface free energy experimentally requires complex measurement machines. The resulting data contains some uncertainties. A good complement are computer simulations with first principle methods (ab initio methods) and the density functional theory [85, 117]. If the surface free energy is known, then the Wulff construction shown on the polar plot will show the equilibrium shape of the crystal. Finally, this equilibrium shape can be used as a input parameter for phase-field simulations.

A strong cubic anisotropy is assumed, such that in 2D a square represents the equilibrium shape. The crystal lattice of each grain is represented by a rotation angle, which defines the orientation of the square. At the grain boundary two crystal shapes are considered by intersecting the squares of the adjacent grains with each other and constructing the inner envelope. The corners of the inner envelope are used in equation (3.4) to simulate strong faceted growth. In contrast to [106], the two squares are always centered and no volume constraint is applied. Figure 3.6 describes the construction of [94].

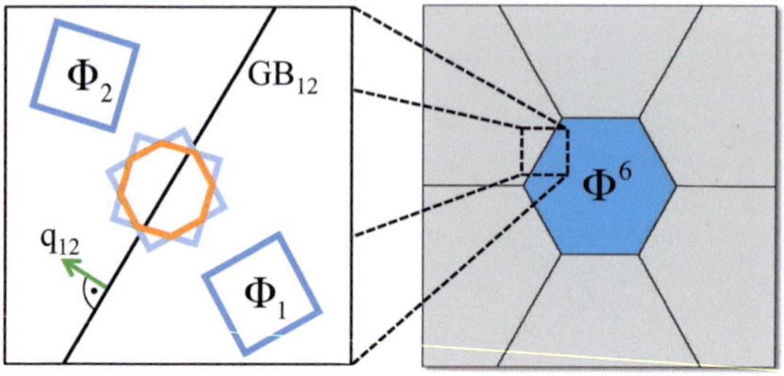

Figure 3.6.: Construction of the intersected cubic anisotropy at the grain boundary. The blue squares represent the cubic crystal shape and lattice orientation. The orange square is the inner envelope of two intersecting squares at the grain boundary. The green arrow is the generalized gradient vector q_{12}.

The gradient vector q_{12} is perpendicular to the grain boundary and determines the grain boundary energy with the inner envelope, according to equation (3.4). The grain boundary energy is not only determined by the crystal lattice orientations but also by the grain boundary orientation. To provide a sharper distinction from the Read–Shockley model, figure 3.7 shows the energy of two zero degree rotated and two 45 degree rotated squares over the grain boundary orientation (q_{ij}).

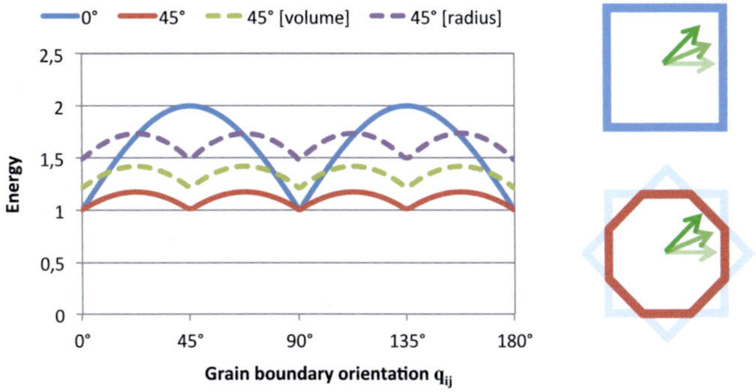

Figure 3.7.: Grain boundary energy for two nearly equal crystal lattice orientations about 0° in blue and the energy of two crystal lattices rotated by 45° in red. The dashed line indicates in green a volume constraint as in [106] or in violet a radius constraint of the 45° rotated lattices. The blue square and the red octagon are the inner envelopes for 0° and 45° rotated lattices.

Parameter	Value
Domain size	400×400
Boundary condition	Dirichlet condition after Voronoi filling
Time steps	200 000
Time step width	0.064
Interface width ε	10
Higher order parameter $\gamma_{ij\delta}$	10
Δx	1

Table 3.1.: Simulation parameters of all presented phase-field simulations.

3.3. Simulation settings

Two different simulation setups are selected to analyze the behavior of grain boundaries. The first setup consists of three grains with one triple junction in the center to observe the movement of the triple junction, according to Young's equations, with the Read–Shockley model. The second setup is used to compare the growth rates of the Read–Shockley model with the growth rates of the faceted grain boundary energy model for n-sided grains with $4 \leq n \leq 8$.

The simulation settings are chosen to meet the phase-field conditions and to minimize the influence of the domain boundaries. Test simulations with a 400×400 cells and a 800×800 cells domain are computed. The smaller 400×400 domain is chosen because the influence of the domain boundary is negligible and the computational effort is smaller. All phase-field simulations are conducted with the same set of parameters, except the crystal lattice orientation and the anisotropy model. The total number of time steps is 200 000 and the simulation state is stored every 100 000th time step, such that 2 snapshots in total are stored. After the first 100 time steps, the simulation state is saved to compute the growth rates with a relaxed phase-field interface. Table 3.1 contains the list of the used phase-field parameters.

Both setups are based on the Voronoi algorithm to create regular grain structures. The first setup is called triple junction setup to analyze the movement of a triple junction for the semi anisotropic Read–Shockley model. Three Voronoi points are placed regularly on a circle around the domain center. The second setup is called n-neighbor setup, where n points are placed regularly around one central Voronoi point. The radius of the circle for both setups is 150 cells. The Voronoi construction for the triple junction setup is depicted in figure 3.8a and the setup for the n-neighbor setup with $n = 5$ is depicted in figure 3.8b. The scaling factor a is used to create grains with a defined surface area for chapter 4. Growth rate distributions.

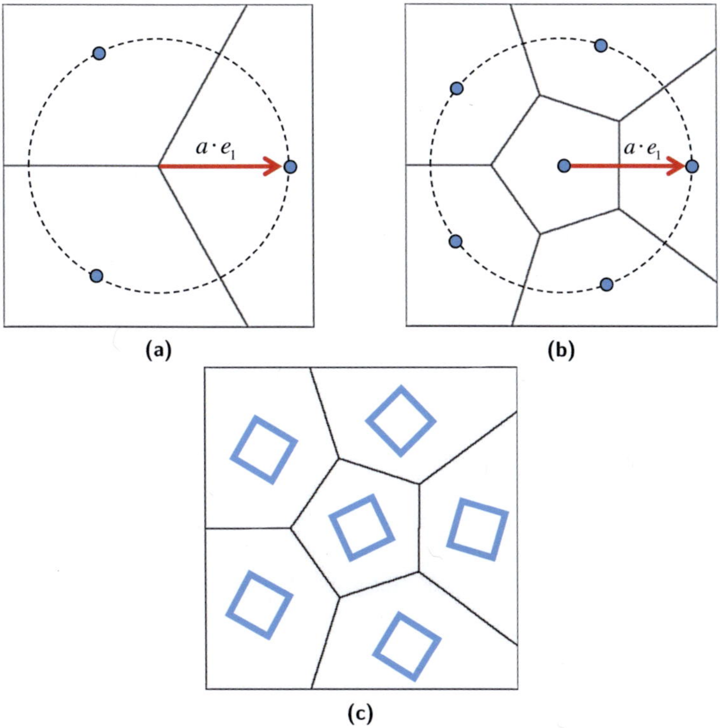

Figure 3.8.: Regular grain structures are constructed by placing Voronoi points regularly on a circle. (a) Three points are used to construct a triple junction in the center. (b) Five Voronoi points are placed around a central Voronoi point to construct a regular grain with 5 neighbors. (c) Randomly rotated crystal lattices are assigned to each grain.

To investigate the grain growth behavior three different simulation series, **triple junction movement**, **isotropic grain growth** and **anisotropic grain growth**, are considered. The first series uses the triple junction setup and the Read–Shockley model. The second series is based on the n-neighbor setup with isotropic grain boundary energies also based on the Read–Shockley model. The last series is based on the n-neighbor setup with anisotropic Read–Shockley and faceted grain boundary energies. Using the Read–Shockley model for isotropic simulations sounds ambiguous. The answer to the question, which energies should be used for isotropic simulations, solves the ambiguity. All grain boundary energies of a simulation are set equally based on the Read–Shockley model with e.g. $7°$ or $14°$.

Triple junction movement
A single triple junction, as depicted in figure 3.8a, is ideal to investigate the influence of different grain boundary energies on a triple junction. For simplicity and to keep the parameter space small, only low (L) and high (H) angle grain boundaries are considered. All eight combinations of high and low angle grain boundaries (LLL, LLH, LHL, HLL, LHH, HLH, HHL, HHH) are simulated with the triple junction setup. High angle grain boundaries correspond to the Read–Shockley model at an angle of $22°$. For the low angle grain boundary, the range from $1°$ to $22°$ is scanned in $1°$ steps.

Isotropic grain growth
The n-neighbor setup is used for two different simulation series. The isotropic series is simulated with isotropic grain boundaries using the same approach as for the low angle grain boundaries. The isotropic energies are based on the Read–Shockley model and are scanned from $1°$ to $22°$ in $1°$ steps.

Anisotropic grain growth
For the anisotropic series, 100 random sets of crystal orientations are drawn. A set consists of nine orientations because the highest number of grains in the n-neighbor setup is nine. Each regular n-sided grain setting is simulated with the Read–Shockley and the faceted grain boundary energy models with 100 sets of orientations, such that this series consists of $5 \times 100 \times 2 = 1\,000$ simulations. In contrast to the first two series, each grain has a crystal lattice orientation which determines the grain boundary misorientation to

reproduce realistic grain boundary misorientation topologies. Figure 3.8c shows the random crystal orientations for one set of orientations with a regular five-sided grain.

3.4. Triple junction movement results

To be able to read and interpret the triple junction simulations, it is helpful to bear Young's equations in mind. Young's equation considers forces tangential to the grain boundaries on the triple junction. High angle grain boundaries are the strongest forces and the force of low angle grain boundaries varies from almost zero to high angle grain boundaries, according to the Read–Shockley model. Figure 3.9 depicts the position of the triple junction after 100 000 time steps for all triple junction simulations. According to figure 3.9, L and H are at the positions [I,II,III]. Each combination of low and high angle grain boundaries e. g. all LLH simulations are marked with the same symbols.

The triple junction for simulations with one high and two low angle grain boundaries follow the grain boundary because two low angle grain boundaries cannot hold the strong high angle grain boundary back. The triple junction for simulations with one low and two high angle grain boundaries move slower in the opposite direction of the low angle grain boundary because two H grain boundaries have to pull the triple junction. This is caused by the geometrical relation of the grain boundaries. LLL and HHH combinations are not plotted because no movement is observed. The most outer position of triple junctions are reached by low angle grain boundaries with 1° because the resulting force vector length is nearly zero and high angle grain boundaries can pull without a retention force. A bigger retention force keeps the triple junction closer to the center.

3.5. Isotropic grain growth results

Three simulation snapshots are stored for each simulation of the isotropic n-sided grain series to compute the growth rates. The first snapshot is taken after time step 100, the second after time step 100 000 and the last after time step 200 000. The last snapshot is not usable, because the central 4-sided grain shrinks too fast and vanishes during snapshot two and three, such that the growth rates are computed between time step 100 and 100 000.

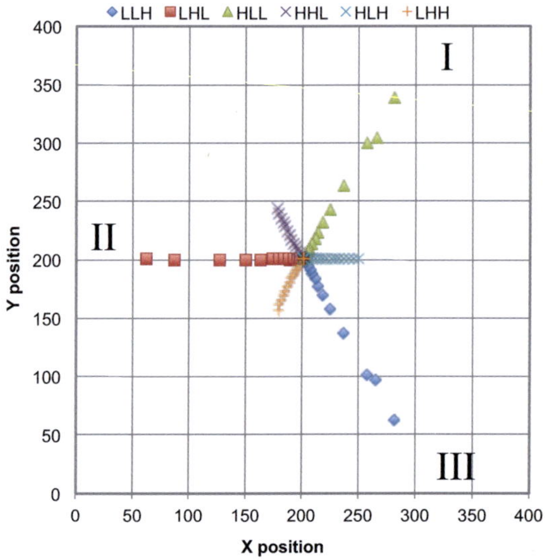

Figure 3.9.: Simulation domain with triple junction positions for combinations of H and L grain boundaries at time step 100 000. The black lines show the initial state. The order of L and H corresponds to the grain boundaries [I,II,III]. The movement of the triple junction corresponds to Young's equations (forces tangential to the grain boundary).

The isotropic growth rates of the simulations are compared with the analytical Neumann–Mullins relation

$$\frac{\mathrm{d}A}{\mathrm{d}t} = -2\pi M\gamma(1 - \frac{1}{6}n), \tag{3.12}$$

where the area change $\frac{\mathrm{d}A}{\mathrm{d}t}$ is determined by the grain boundary energy γ, the mobility M and the number of neighbors n. The measured and analytical growth rates for each regular n-sided grain setting is plotted over the misorientation angle in figure 3.10. The simulations follow the analytical solution with a deviation. This deviation is related to the not properly established phase-field interface at time step 100. Another snapshot timing related problem are higher deviations for angles smaller than 6°. The cusp for the 8-sided grain at 4° cannot be explained by means of snapshot timings or parameter settings. Because this irregularity is observed only in one simulation, I assume a software or hardware issue.

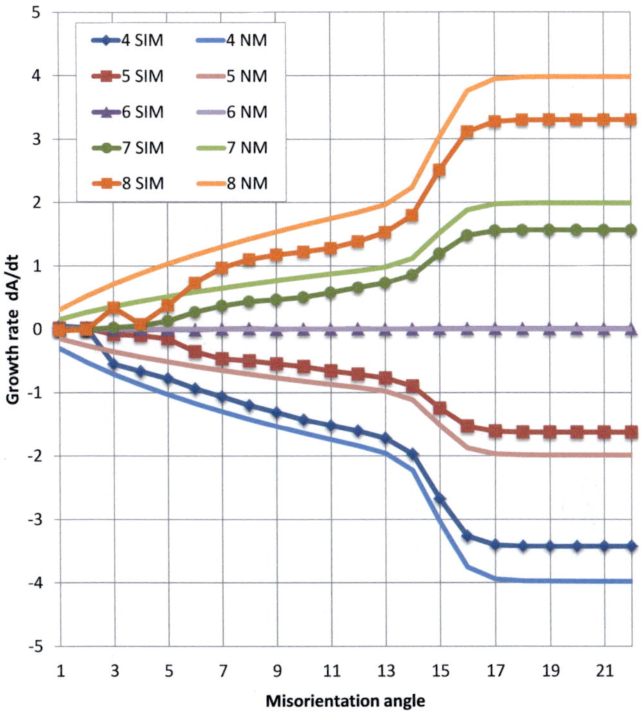

Figure 3.10.: The isotropic simulations follow the Neumann–Mullins relation with a small offset. For misorientation angles smaller than 5°, the simulations deviate from the Neumann–Mullins relation, which may be caused by a measurement snapshot taken too early.

3.6. Anisotropic grain growth results

The Read–Shockley as well as the facetted anisotropy model follow the isotropic Neumann–Mullins relation in average. The maximum, average and minimum growth rates are depicted in figure 3.11 for the Read–Shockley model and for the faceted anisotropy model. The average growth rates form a line with positive slope and a zero growth rate at $n = 6$. The isotropic system with only high angle grain boundaries has the steepest slope followed by the Read–Shockley model and the faceted anisotropy model with the flattest slope. The maximum and minimum lines of the Read–Shockley model are parallel to the average values and shifted. The minimum line of the Read–Shockley model is flatter for $4 \leq n \leq 6$ than for $n > 6$. The fastest shrinking 4-sided grain and the fastest growing 8-sided grain have nearly the same growth rates as the isotropic simulations.

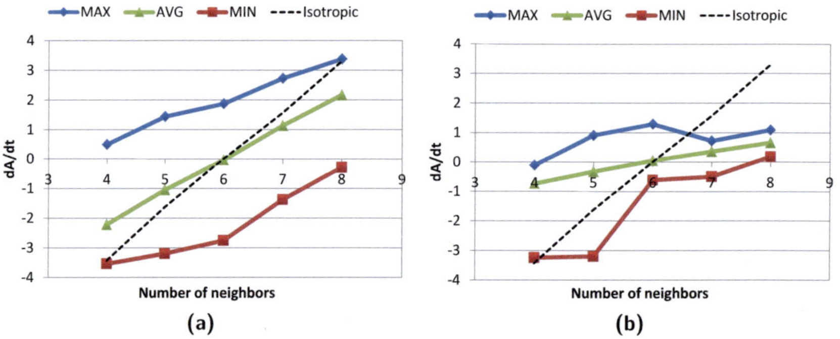

Figure 3.11.: Minimum, average and maximum growth rates of the (a) Read–Shockley and (b) faceted anisotropy model simulations for each 100 n-sided grain simulations.

In contrast to the Read–Shockley model, the maximum and minimum values of the faceted anisotropy model do not form straight lines. A significant step for the minimum growth rate from $n = 5$ to $n = 6$ is produced by outliers, which show a significant shrinkage. Outliers in the opposite direction lead to the bump at $4 \leq n \leq 6$. No maximum and minimum outliers are present for $n > 6$. A separate description of the outliers follows in the next section.

Isotropic		Read–Shockley				Faceted			
n	∅	max	∅	min	σ	max	∅	min	σ
4	-3,43	0,49	-2,22	-3,54	1,14	-0,11	-0,74	-3,25	0,42
5	-1,63	1,44	-1,06	-3,20	1,12	0,90	-0,33	-3,20	0,36
6	0,01	1,87	-0,04	-2,76	1,04	1,28	0,04	-0,61	0,19
7	1,56	2,72	1,13	-1,38	0,83	0,72	0,36	-0,50	0,13
8	3,30	3,37	2,16	-0,29	0,82	1,09	0,66	0,18	0,12

Table 3.2.: Maximum (max), average (∅), minimum (min) and standard deviation (σ) of the growth rates for the Read–Shockley and faceted anisotropy simulations for each n-sided grain setup.

Statistical data in table 3.2 show the absolute values of figure 3.11 with the standard deviation σ for each n-sided grain configuration. The average growth rates and the standard deviation of the faceted anisotropy are significantly smaller than for the Read–Shockley model. A negative average growth rate for setups with $n < 6$, almost no growth for $n = 6$ and a positive average growth rate for setups with $n > 6$ is present for all surface energy models. Another common property is the reduction of the standard deviation with increasing neighbor count for both anisotropy models. Outliers with the faceted anisotropy are indicated by the large deviation between the average values and the standard deviations. This deviation is significantly smaller for the Read–Shockley simulations.

3.7. Fast growth with faceted anisotropy

Some growth rates of the simulations with faceted anisotropy deviate strongly from the average value and are not in close relation to the standard deviation. A growth rate is defined as an outlier, if the growth rate is outside the range of $\pm 4\sigma$ around the average growth rate. Five simulations (1%) out of 500 are identified as outliers with this definition. Two with a high positive growth rate and three with a high negative growth rate.

The high growth rate of the 6-sided grain is a good example to show the grain evolution. The starting conditions and the grain numbers are depicted in figure 3.12a. Figure 3.12b shows that grain Φ_0 has grown into grain Φ_2 and Φ_3 after 200 000 time steps, while the grains Φ_5 and Φ_6 are almost unaffected. Only the central grain Φ_0 and the grains Φ_1, Φ_2 and Φ_3 seem to be necessary for the fast growth of the grain boundary GB_{02}. To increase the growth rate a separate simulation is conducted, where the symmetry of the 6-sided grain is used to increase the growth rate. The orientations of the grains Φ_1, Φ_2 and Φ_3 are copied to the grains Φ_4, Φ_5 and Φ_6, as depicted in figure 3.12c. A domain with 800×800 cells and with 2 million time steps is used. The evolution in figure 3.12c shows a faster growth at the beginning. The same behavior is observed for shrinking grains. One side is growing very fast into the central grain and the symmetry can be used to increase the negative growth rate, as depicted in figure 3.13.

According to the performed simulation series, the probability of one outlier with one fast moving grain boundary is assumed to be 1%. The probability of a fast growing or fast shrinking grain, as with the modified 6-sided grain setup, is estimated to be $\frac{1}{n \cdot 10\,000}$. The probability of one fast moving grain boundary is $\frac{1}{100}$ and the probability of another fast moving grain boundary at the opposite position with a e. g. 6-sided grain is $\frac{1}{100} \cdot \frac{1}{6}$, such that the resulting probability is $\frac{1}{60\,000}$.

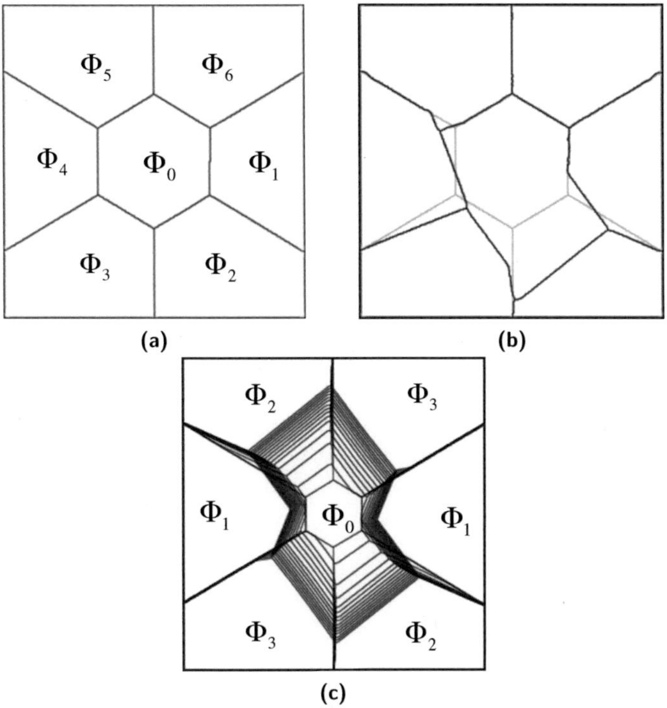

Figure 3.12.: A modification of the fastest growing 6-sided grain setup increases the growth speed further. (a) The 6-sided grain setup with crystal lattices from Φ_0 to Φ_1. (b) Grain zero grows into grain two, while the grains five and six are unaffected. (c) The growth rate can be increased by using the crystal lattices Φ_1, Φ_2 and Φ_3 and the 6-sided symmetry of the grain.

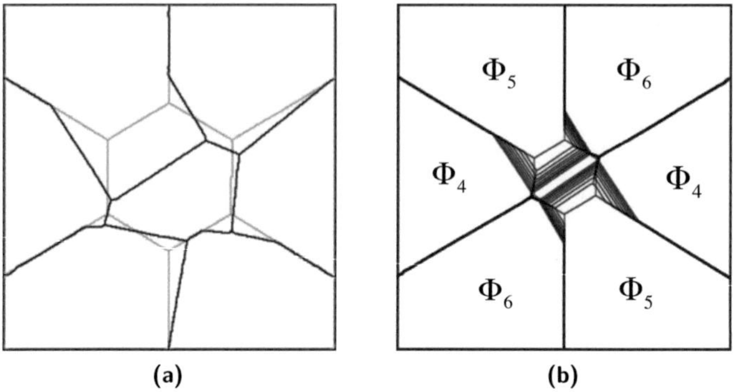

Figure 3.13.: A modification of the fastest shrinking 6-sided grain setup increases shrinkage speed further. (a) The 6-sided grain setup with crystal lattices from Φ_0 to Φ_1. (b) Grain five grows into grain zero. (c) The shrinkage rate can be increased by using the crystal lattices Φ_4, Φ_5 and Φ_6 and the 6-sided symmetry of the grain.

3.8. Summary

The triple junction movement series with the semi-anisotropic Read–Shockley model shows a sufficient fulfillment of Young's law by the chosen simulation settings. A complete validation would either require a mathematical proof of the phase-field model or a representative scan of the parameter space.

The isotropic grain growth series does not reveal a new scientific insight but the results confirm the model validity for isotropic microstructures and they serve as a reference for comparisons.

A careful interpretation of the results is:

The average growth rate of regular n-sided grains with $4 \leq n \leq 8$ and random crystal lattice distribution behaves isotropically for the Read–Shockley and the strongly faceted grain boundary anisotropy model.

This conclusion has to be taken with care because a representative behavior of the random setting is assumed. In fact, 1 000 simulations are a very small sample of the parameter space and far away from a statistical representative set of simulation settings. If each crystal lattice of a e. g. 6-sided grain is scanned in 1° steps, the total amount of simulation settings is 90^6, such that follow up questions are:

- Is the conclusion true for a statistical representative set of simulation settings?
- Is the simulation of a statistical representative set possible with the current computation facilities?
- What do the growth rate distributions look like?
- What are the settings for the fastest growth/shrinkage?

By means of analysis and simulations, Kazaryan, Patton, Dregia and Wang describe in [49], that the average growth rates of a grain structure with Read–Shockley grain boundary energy will behave isotropically with random lattice orientations. The same isotropic behavior with the Read–Shockley model is supported in the study of Elsey, Esedoḡlu and Smereka in [22].

Surprisingly, no systematic study of growth rates for faceted grain growth is found. An analysis of faceted anisotropy may not be easy to achieve but straightforward simulation studies are able to point the direction towards interesting findings. The observed growth rate outliers with faceted anisotropy are an interesting finding, which will require a thorough study to exclude model or numerical side effects.

Another result of anisotropic simulation series is, that the Read–Shockley and the faceted grain boundary energy model reduce the growth rate standard deviation with increasing neighbor count. One explanation is, the more neighbors a grain has, the smaller the influence of a single neighbor on the growth rate.

The next chapter tackles the open questions and presents a straightforward way of investigating the effect of anisotropy on grain structures.

4. Growth rate distributions

The Neumann–Mullins relation holds for randomly oriented 2D grains with Read–Shockley-like surface energy on average. However, how are the growth rates distributed for a given topological class of n-sided regular grains, and does abnormal grain growth occur?

Abnormal grain growth is one of the phenomena which hides its physical sources in a huge parameter space. It is well understood, that abnormal grain growth occurs, if one grain has a growth advantage due to its grain boundary energy or mobility property, compared to the residual grains in the polycrystalline microstructure [99, 98, 44, 45]. In a review [97] about grain growth and recrystallization, Rollett, Brahme and Roberts imply a weak relation between grain boundary energy anisotropy and abnormal grain growth. In contrast, studies of perovskite materials [10, 15, 59] show that faceted growth of grains is necessary for abnormal grain growth.

A large computational study illustrates, that abnormal grain growth is strongly related to anisotropic grain boundary properties, and that computation power of supercomputers and powerful simulation tools can be employed to expand the various properties and correlations depending on the crystal orientation. To our knowledge, no previous attempt to obtain growth rate distributions and to relate abnormal grain growth to anisotropy has been invested to scan the parameter space with high resolution.

In this study, the well-known Read–Shockley model [93] is used to investigate the growth rates of regular 2D n-sided grains. The Read–Shockley model takes the misorientation angle of the crystal lattices as a parameter for the grain boundary energy, such that the parameter space consists of the crystal lattice orientations for each neighbor of the n-sided grain within $3 \leq n \leq 8$. Each lattice has a cubic symmetry, such that each lattice is rotated in the range from $-45°$ to $45°$ with small step widths in the 2D plane to compute the growth rate for each configuration.

Unfortunately, precise 2D grain structure simulations with the phase-field method are computationally expensive, compared to the 2D vertex model, as e. g. proposed by Kawasaki, Nagain and Nakashima in [48]. According to Kawasaki et al., the vertex model does not capture curvature like the vertex model of Weygand, Bréchet and Lépinoux in [126], but it is computationally inexpensive. The computational costs play a major role in the derivation of statistically meaningful statements and in computing as many growth rates as possible. A simplified vertex model according to Kawasaki et al. is used, which provides the computational efficiency to obtain growth rates for billions of configurations. To validate the triple junction tracking model, the results of 700 phase-field simulations are compared with results of the triple junction tracking model for the same configurations. A sufficiently linear correlation is obtained, which supports the computed growth rate distributions.

4.1. Method

The following approach is the attempt to obtain the growth rates of all lattice configurations for an n-sided grain in 2D with Read–Shockley grain boundary energy. Therefore, each cubic lattice of the n-neighbors is rotated with small step widths in 2D to get as many configurations as possible, and to compute the growth rate for each configuration. Because of the detailed scan of the large parameter space, it is necessary to use a computationally inexpensive vertex model, according to Kawasaki et al.[48]. The simplified vertex model, which is called by the more meaningful name 'triple junction tracking model', is validated with results of phase-field simulations. This section describes both models, the parameters and setups.

4.1.1. Vertex model - Triple junction tracking model

The change in the occupied area of a 2D grain embedded in a grain structure, is related to the movement of its triple junctions. To predict growth or shrinkage in a computationally inexpensive way, the vertex model is based on the Herring equation [39], where forces act on triple junctions. Only forces on triple junctions, tangential to their grain boundaries (grain boundary forces) are considered. The forces are constructed by the surface energies and mobilities of the grain boundaries. To compute the new shape and area of a grain, each triple junction is moved by its resulting force vector. Curvature of grain boundaries is neglected to reduce the computational effort. After computing the area difference of a configuration, the results are stored. A new cycle begins and the same procedure is started with a new lattice configuration.

To algorithmically construct the ideal shape of an n-sided grain Φ_c^n as a regular polygon, triple junctions are placed on the unit circle around the center point C with the unit vector $\boldsymbol{e}_2 = (0, 1)$ by rotating the unit vector by an angle of $\frac{360°}{n}$. For simplicity, a clockwise numbering index i, which is in the set of integers modulo n ($i \in \mathbb{Z}/n\mathbb{Z}$), is introduced for the triple junctions T_i and grains Φ_i. An exemplary construction for $n = 5$ is shown in figure 4.1[4].

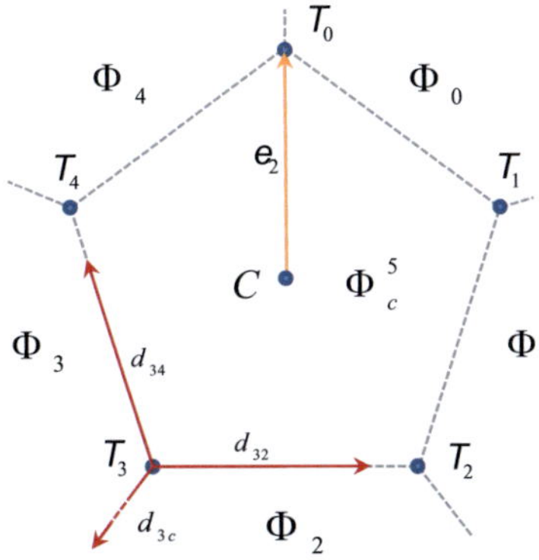

Figure 4.1.: Sample construction of a five-sided grain Φ_c^5 around the center C by rotating the unit vector $\boldsymbol{e}_2$ around the center. Grains surrounding the central grain are numbered clockwise starting with 0. The grain boundary direction vectors $\mathbf{d}_{32}$, $\mathbf{d}_{34}$ and $\mathbf{d}_{3c}$ of the triple junction T_3 are shown by arrows.

[4]IOP Publishing is holding the copyright of the figures in this chapter. The figures are reproduced by permission of IOP Publishing.

Each triple junction T_i has three grain boundary direction vectors, tangential to their grain boundaries with length 1. The direction vector $\mathbf{d}_{i,i-1}$ is pointing to the previous triple junction T_{i-1}, the direction vector $\mathbf{d}_{i,i+1}$ is pointing to the next triple junction T_{i+1}, and the third vector $\mathbf{d}_{i,c}$ is pointing to a virtual triple junction outside grain Φ_c^n.

In this study, both isotropic grain boundary energy as well as the Read–Shockley grain boundary energy are analyzed. In the isotropic case, the grain boundary surface tension between the grains Φ_i and Φ_j $(j \in \mathbb{Z}/n\mathbb{Z})$ is set to $\gamma_{ij} = 1$ and the grain boundary mobility is set to $\tau_{ij} = 0.53$. For the Read–Shockley model, each grain Φ is assigned one orientation angle Θ. The misorientation Θ_{ij} between grains Φ_i and Φ_j with cubic symmetry to compute the surface tension and mobility is formulated according to

The Read–Shockley model is valid for small misorientation angles with $\Theta_{\max}$, commonly between 0° and 45° to adapt to experiments. $\Theta_{\max}$ is set to 15° to model high mobility above 15° for high angle grain boundaries. The ratio $\frac{\gamma_{ij}}{\tau_{ij}}$ is the scaling factor for the direction vectors to define the force vectors acting at the triple junctions, as shown in the equations (4.1) to (4.3).

$$\mathbf{f}_{i,i-1} = \mathbf{d}_{i,i-1} \cdot \frac{\gamma_{c,i-1}}{\tau_{c,i-1}} \tag{4.1}$$

$$\mathbf{f}_{i,c} = \mathbf{d}_{i,c} \cdot \frac{\gamma_{i,i-1}}{\tau_{i,i-1}} \tag{4.2}$$

$$\mathbf{f}_{i,i+1} = \mathbf{d}_{i,i+1} \cdot \frac{\gamma_{c,i}}{\tau_{c,i}} \tag{4.3}$$

$$\mathbf{F}_i = \mathbf{f}_{i,i-1} + \mathbf{f}_{i,c} + \mathbf{f}_{i,i+1} \tag{4.4}$$

All force vectors $\mathbf{f}_{i,j}$ at one triple junction T_i form the resulting force vector $\mathbf{F}_i$, as illustrated in figure 4.2a. With the position of the triple junctions and the force vectors, the new shape and area A' of the grain Φ_c^n is determined by moving all triple junctions with their resulting force vectors, as shown in figure 4.2b.

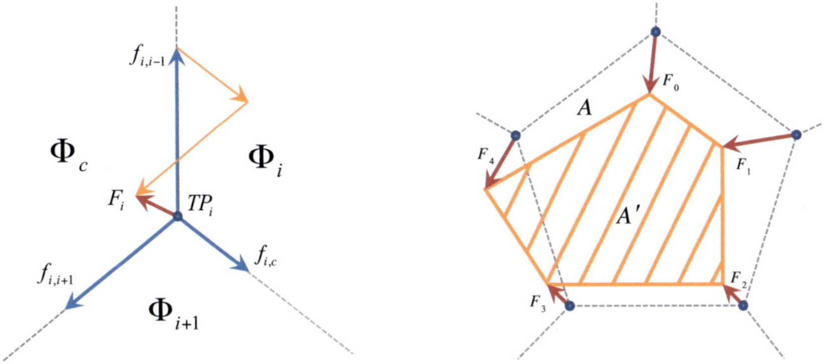

Figure 4.2.: (a) Construction of the resulting force vector $\mathbf{F}_i$ at the triple junction T_i composed of grains Φ_c, Φ_i and Φ_{i+1}. (b) New grain shape and area A' by moving each triple junction T_i with its resulting force vector $\mathbf{F_i}$.

An overlap of two force vectors on one grain boundary is avoided to prevent the formation of self-intersecting grain shapes. Because the maximum length of the force vectors is ≤ 1.9, the initial area A is increased by scaling the unit vector $\boldsymbol{e}_2$, such that each grain boundary is ≥ 4 length units. The approach allows the use of the shoelace formula, to calculate A and A', to obtain the growth rate $dA = A' - A$. Based on the model formulation, a one-step forward prediction with a single snapshot is performed where time scaling is not incorporated. Time scaling is required to introduce a scaling factor of the growth rates, which is equal for all computed dA, as long as self-intersecting grain shapes are avoided.

The implemented program scans the 2D orientation space with a given step width and within a given interval of the crystal lattice orientation. The growth rate of each configuration is stored in a histogram database. Because of the high number of configurations, it is necessary to parallelize the program. The Cray MPICH2 implementation of the message passing interface standard [72] for parallel computation is used. Each configuration can be computed independently, such that the search space is divided in parts, which are assigned to parallel tasks.

4.1.2. Setup

To validate the triple junction tracking model, 100 phase-field simulations with random grain orientations are set up for each topology class from 3 to 9 neighbors, as shown in figure 5.6, such that 700 phase-field simulations are computed in total. Each simulation is conducted with the same set of parameters, except for the crystal lattice orientations. The crystal lattices are rotated in the simulation plane. In order to measure the area change, a snapshot is taken after 100 time steps to assure a proper phase-field interface. After 20 000 time steps, the grain ensembles are evaluated to measure the area change. The fastest growing and shrinking grains determine the simulation time, to avoid vanishing of grains or hitting the boundary. The same grain boundary properties formulated in equation (3.10) and (3.11) of the triple junction tracking model are used for the phase-field simulations. Table 3.1 of the previous chapter contains the list of the used phase-field parameters. The only difference to the simulations in the previous chapter is a reduction of the time steps to 20 000.

As displayed in Fig. 4.3, the Voronoi algorithm is used to construct the starting conditions for the phase-field simulations. One Voronoi point is placed at the center of the domain and the other n Voronoi points are placed along a circle around the center with radius $a \cdot |\mathbf{e}_1|$. For a grain with 5 neighbors, the resulting structure can be seen in figure 4.3. The area of the central grain is set to 20 000 cells for all simulations, which is determined iteratively with the Newton method by varying the scaling factor a. To validate the triple junction tracking model, all 700 phase-field configurations are also computed with the triple junction tracking model.

In order to obtain growth rate distributions with 3 to 8 neighbors, including the Read–Shockley model, it is necessary to simulate the growth behavior for many different configurations in order to achieve enough data points for a representative statistics of misorientations. The setup of a 2D grain with eight neighbors, for instance, has eight degrees of freedom, because each neighbor grain has one lattice orientation in 2D. Depending on the neighbor count, each lattice orientation is scanned from $-45°$ to $45°$ with different step widths to obtain the results in reasonable time. Table 4.1 contains the step width for each topological class.

As mentioned in section 4.1.1, the overlapping of resulting force vectors is avoided by choosing an area of 78.0 for all configurations to obtain edge lengths of at least 4. The area of a regular polygon is again determined iteratively with the Newton method by scaling the unit vector $\mathbf{e}_2$ of the polygon construction in figure 4.1.

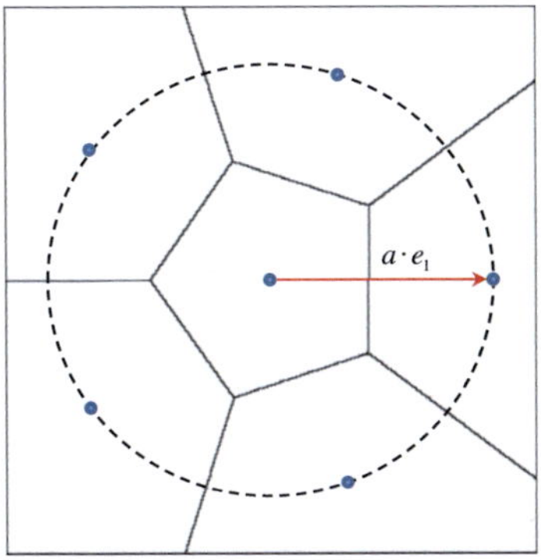

Figure 4.3.: Construction of the starting conditions for the phase-field simulations with the Voronoi algorithm. Five Voronoi points are placed regularly on the circle with a radius of $a \cdot |\mathbf{e}_1|$ around the central Voronoi point, to obtain a grain with 5 neighbors.

4.1.3. Parallelization

Parallelization is necessary to compute a statistically meaningful number of configurations in reasonable time. The parallelization scheme of the phase-field implementation is based on domain decomposition, which is presented in chapter 5. The parallelization scheme for the triple junction tracking model is based on the decomposition of the parameter space. Each processor computes a part of the parameter space.

A regular 3-sided grain with a step width of 0.1° is assumed to explain the parallelization scheme with 64 processors. The total number of configurations is $N = (\frac{90}{0.1} + 1)^3 = 731\,432\,701$. Every process has to compute $\lfloor \frac{N}{64} \rfloor$ configurations. Because the number of iterations in not dividable without a remainder, every process with an id smaller than $N \mod 64$ gets one additional configuration to compute. The final step is to compute the starting configuration to compute the growth rates and iteratively rotate the crystal lattices. Each process rotates the crystal lattices step by step until each process reaches the final configuration of its part of the parameter space.

n	Step width	Config.	CPUs	Seconds	$\frac{\text{Config.}}{\text{Second}\cdot\text{CPU}}$
3	0.10°	731 432 701	1 024	40	17 857
4	0.25°	16 983 563 041	1 024	1 177	14 091
5	0.75°	25 937 424 601	2 048	1 084	11 683
6	1.50°	51 520 374 361	2 048	2 533	9 931
7	2.50°	94 931 877 133	2 048	4 864	9 530
8	4.00°	78 310 985 281	2 048	4 628	8 262

Table 4.1.: Setup of the triple junction tracking computation runs on the HERMIT supercomputer in Stuttgart (Germany). The more neighbors are investigated, the bigger the step width because of the exponential growth of the configuration count. An uneven load distribution can be derived from the number of configurations one CPU can compute in one second.

4.2. Results

The timings of the approx. 268 billion configurations computed with the parallel triple junction tracking implementation, are shown in table 4.1. Even with a computationally inexpensive model, it takes four hours with over 1 024 processors to compute approx. 268 billion configurations, resulting in a sequential runtime of about 11 months. Each of the 700 phase-field simulations, to validate the triple junction tracking model, is computed with 16 processors and a runtime of about 20 seconds.

Validation
A sufficient correlation of the growth rates between the phase-field simulations and the triple junction tracking computation results has to be achieved, to support the validity of the triple junction tracking results. In order to validate the triple junction tracking model with the phase-field simulations, the growth rates of the 700 phase-field simulations are related to the same configurations for the triple junction tracking computations. Figure 4.4 shows a linear correlation of the growth rates with a Pearson product-moment correlation coefficient of 0.96 between the phase-field simulations and the triple junction tracking computations. A correlation coefficient near 1.0 expresses a strong correlation between the results of the two models, so that the results of the triple junction tracking model are considered as sufficiently valid.

Growth rate distributions
The results of each supercomputer run are stored in a histogram with 550 buckets between the growth rates -60 and $+50$ to save memory. All histograms are normalized and combined in figure 4.5. They show the movement of the distributions from negative growth rates with less neighbors to positive growth rates with more neighbors. Beside the movement of the distributions, the shape of the distributions is not symmetric and, with more neighbors, it changes towards a normal distribution with decreasing variance.

One striking property are the peaks, which result from our Read–Shockley anisotropy model. The model behaves isotropically above a misorientation of $17°$, with constant grain boundary energy of about 1.9. Therefore, about 62% of the misorientation range from $0°$ to $45°$ behaves isotropically. This proportion leads to the peaks in all grain growth distributions.

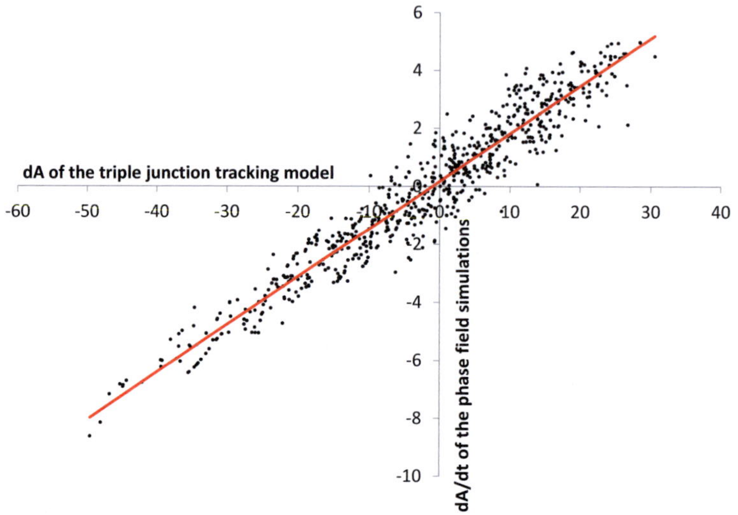

Figure 4.4.: A linear correlation (red regression line) of the growth rates between the 700 phase-field simulations and the corresponding triple junction tracking computations exists. The Pearson product-moment correlation coefficient is about 0.96.

n	Isotropic dA	Average dA	Shrinking	Growing
3	-25.49	-20.08	98%	2%
4	-18.42	-14.38	88%	12%
5	-8.82	-7.23	74%	26%
6	0.00	-0.07	47%	53%
7	7.51	5.80	24%	76%
8	13.81	12.54	6%	94%

Table 4.2.: The isotropic growth rates of the triple junction tracking model correspond to the peaks in figure 4.5. The average values of the Read–Shockley model behave isotropically with a deviation from the pure isotropic case. The last two columns show that not all configurations with three, four and five neighbors shrink and that not all grains with seven and eight neighbors grow.

For example, a triple junction tracking computation with isotropic grain boundary energy and 3 neighbors, as shown in table 4.2, provides the isotropic growth rate of -25.49, which corresponds to the position of the peak in figure 4.5a. These peaks originate from the isotropic part of the anisotropy model where all grain boundary misorientations are above $17°$. With an increasing neighbor count, the number of such combinations decreases, hence the peaks shrink.

Another distinctive feature are the cusps, which reduce and finally disappear for an increasing neighbor count. The growth rate distribution for three neighbors in figure 4.5a shows two small and four large cusps. Starting from left to right, two small cusps are at the leftmost position followed by four larger cusps. In figure 4.5b with four neighbors, the number of large cusps is five. Figure 4.5c corresponds to five neighbors and indicates a sixth cusp at the rightmost position, which gets superimposed by noise. The number of large cusps is by one greater than the number of neighbors. For more than five neighbors, the cusps are not clearly visible because they move too close to each other and adapt to a normal distribution. The cusps indicate groups of configurations with similar growth rates.

Average behavior

Another representation of the results, with a whisker box plot in figure 4.6, illustrates the movement of the mean value from negative to positive growth rates, and the decreasing variance with increasing neighbor count. The mean growth rate follows the Neumann–Mullins relation. The range from minimum to maximum growth rates between even and uneven neighbor counts shows oscillations, which seem to originate from the symmetry of the grains. These oscillations of the minimum and maximum growth rates can be noticed but they need to be considered with care, because of the simplicity of the triple junction tracking model.

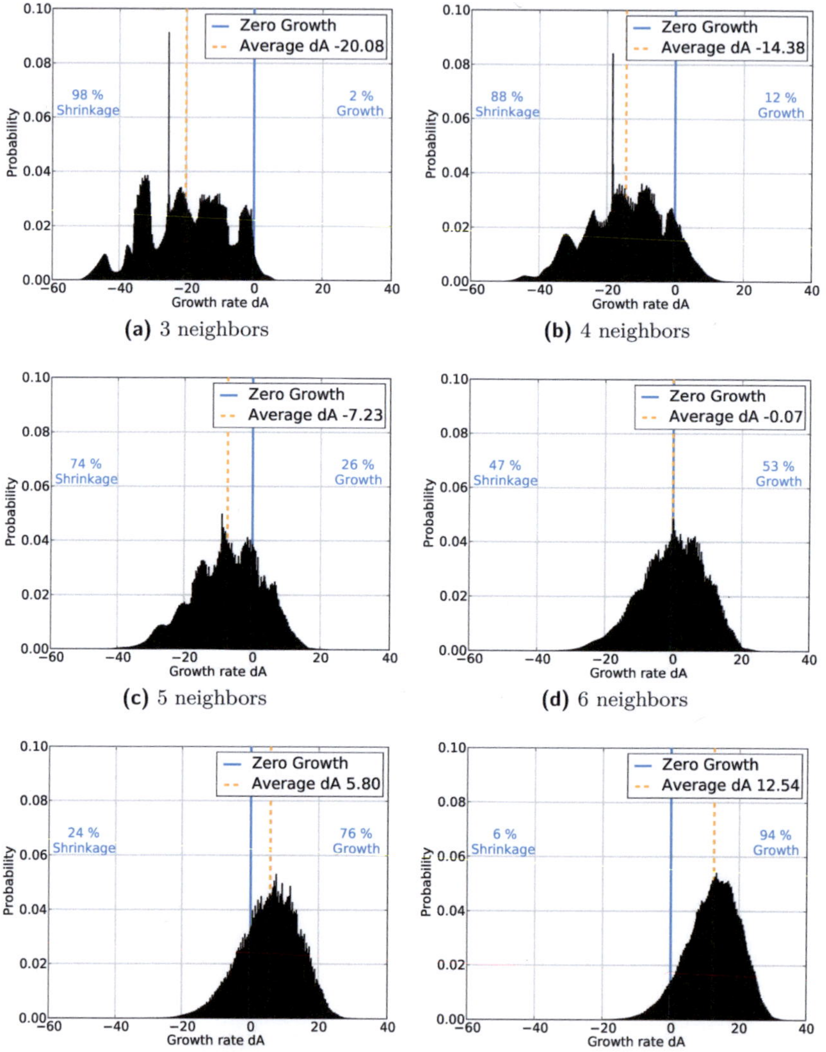

(a) 3 neighbors

(b) 4 neighbors

(c) 5 neighbors

(d) 6 neighbors

(e) 7 neighbors

(f) 8 neighbors

Figure 4.5.: Normalized growth rate distributions for 3 to 8 neighbors of the triple junction tracking runs on the HERMIT supercomputer. For an increasing number of neighbors, the distributions shift to a positive growth regime, the isotropic peak aligns with the average growth rate, the variance decreases, and the shape transforms towards a normal distribution.

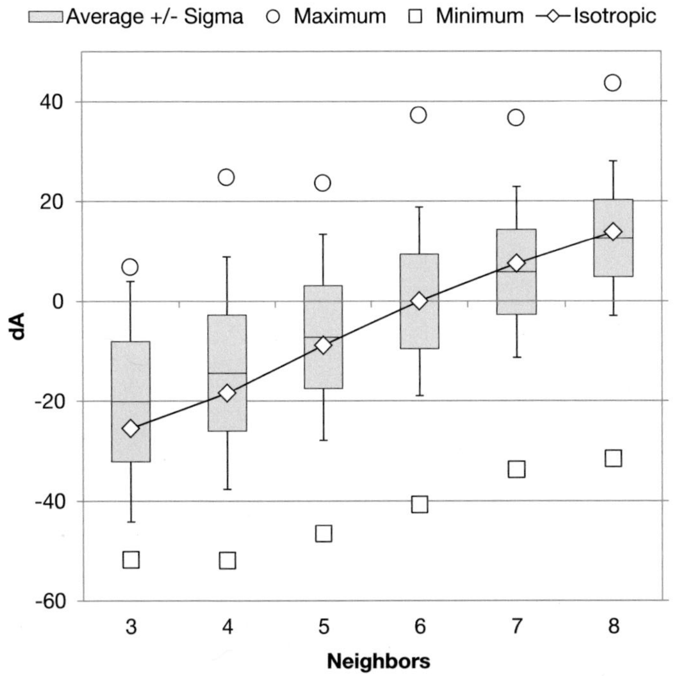

Figure 4.6.: Representation of growth distributions with isotropic growth rates (open diamonds), $\pm$ variance σ (grey box), $\pm 2\sigma$ (whisker) and extreme values (open circles/squares). The average growth rate follows the Neumann–Mullins relation and moves from negative to positive values, with increasing neighbor count and decreasing variance.

4.3. Discussion

Even with the use of a supercomputer, the view of the complexity of the grain behavior with anisotropic grain boundary properties is limited because of the immense parameter space. A combined approach of a regular scan and probabilistic selection of configurations will be necessary to investigate the growth rates for n-sided grains with more than 10 or 20 sides to answer the question of whether the variance of the growth rate distributions decreases to zero or saturates for an increasing neighbor count.

As discussed in the following, related studies are found that support the results and help to understand the effects of the Read–Shockley anisotropy model. The obtained growth rate distributions support the findings described by Kazaryan, Patton, Dregia and Wang in [49]. They showed that the growth rates of a grain structure with Read-Shockley grain boundary energy will behave isotropically for grain structures with random orientations. The same isotropic behavior is supported in the study of Elsey, Esedoglu and Smereka in [22]. In contrast to their work, a Θ_{max} of $15°$ instead of $30°$ is used in equation (3.10), which increases the isotropic peaks but does not change the isotropic character for random orientations.

Ignoring the grain boundary curvature and different mobilities for triple junctions is a simplification that affects the growth behavior, as shown by Gottstein, Ma and Shvindlerman in [31]. They showed the influence of triple junctions with limited mobility on grain structures. Limited triple junction mobility introduces anisotropy, hence the dynamics is changed. A recent study of triple junction mobility with the phase-field method by Johnson and Voorhees in [47], explains the influence of triple junction mobility at low temperature and small length scales on grain growth. Incorporating curvature or a triple junction mobility model extends the validity of the results and the parameter space. Simultaneously, the computational effort and operating processors increase by some orders of magnitude.

4.4. Conclusion

For the first time, the growth rate distributions of the Read–Shockley model are determined by statistically representative computations for regular 2D grains with 3 to 8 sides. This computationally intense task of computing the growth rates of about 268 billion configurations, could only be accomplished in reasonable time with a simplified vertex model, according to Kawasaki et al. and a powerful computation facility. In order to perform such studies, proven simulation methods such as the phase-field method are also required to validate the idealized vertex models. The study of the presented high resolution scan is performed in two dimensions, applying a Read–Shockley model. The results reveal some properties, which are summarized as follows:

- Lattice configurations can be detected with the Read–Shockley model, which behave contrary to the Neumann–Mullins relation,
- the Read–Shockley model follows the isotropic Neumann–Mullins relation for 2D grain structures with random lattice orientations,
- the growth rate distributions with the Read–Shockley model are not symmetric and adapt a normal distribution for 3 to 8-sided regular 2D grains,
- abnormal grain growth behavior can not be explained with the Read–Shockley model for 3 to 8-sided regular 2D grains

A thorough study of a proposed anisotropy model is essential to capture all of its effects and limitations, which can lead to surprising behavior, like the isotropic peaks in the growth rate distributions of figure 4.5. Going beyond two-dimensional simulations into the more realistic 3D space and extending the limited one-parameter Read–Shockley model, which is only misorientation-angle-dependent, will increase the effort to reveal phenomena caused by anisotropy. Generalizations of the current work will access to explore new scientific findings in the field of anisotropic grain growth as e. g. experimentally observed phenomena such as abnormal grain growth, which are strongly related to grain boundary anisotropy.

5. Performance Analysis

Performance plays a significant role in the framework of computational science to simulate phenomena within an appropriate time frame or with an appropriate level of detail. It is not always the case to consider performance concerns within the scientific work, because the development of a model and the validation and investigations with the model are of greater interest and importance. The quality of the results a scientific application generates is the main objective of scientific work. In some cases, it is necessary to consider performance as one of the scientific goals to simulate large systems with a high level of detail. In this work, it is necessary to assure high performance to simulate recrystallization in 3D with a large microstructure to obtain the results. This chapter covers the performance analysis of the Pace3D (Parallel algorithms for crystal evolution in 3D) solver regarding the parallel runtime behavior.

The software package Pace3D contains elaborated tools as well as a versatile and modular implementation of several phase-field models. It solves phase-field-type models and provides a compendium of methods to study microstructure formations in multiphase and multicomponent material systems taking into account the influences of heat and mass transfer, fluid flow as well as mechanical forces. Pace3D is not restricted to simulations of microstructures, such that simulations of foam, bubble structures and particles like blood vessels can be performed. It is programmed in C/C++ and contains parallelism, adaptive numerical solving schemes, a concise data structure management and concepts for an integrated coupling of modeling and data analysis technologies.

Large simulations in 3D can take months or even years of computation on a regular workstation. In order to obtain results in days or weeks, it is necessary to compute the simulations in parallel. Pace3D contains many optimizations to execute 3D simulations and it is already parallelized with the Message Passing Interface standard (short MPI) defined in [72]. MPI is designed for distributed memory computation without a global shared memory, such that messages have to be send and received explicitly. The parallelization with MPI enables to simulate on supercomputers and high performance cluster (short HPC) systems.

A supercomputer is a system, which is one of the fastest 500 systems in the world [74]. To capture more systems and to ease reading, it is referred to as HPC rather than supercomputer. Different architectures exist, but they share common properties. HPC systems consist of computation nodes, which are connected with a high performance network. The computation nodes are merely workstations with more and faster memory, as well as powerful CPUs. An ordinary workstation main board has one CPU socket, in contrast to a HPC computation node board with several CPU sockets. The other main difference despite memory and CPU is the network. A high performance network is not only very fast, its topology is optimized for efficient communication. A fat tree topology [61] for example is very common. About 42% of the top 500 computers [74] use infiniband with a fat tree topology. The advantage is a fast network with a small diameter. The diameter of a network is the longest shortest path and it is favorable to keep the diameter small. Another advanced topology is the torus network [133], which also has a small diameter, low cost and good scaling properties.

MPI is a well-suited standard for computational science on distributed memory computers like HPCs. Specialized implementations considering the existing network topology and network hardware allow to achieve high performance. On the other hand, it is not mandatory to achieve good scalability by using MPI carelessly. All of the top 500 systems have more than 8 000 CPUs and an appropriate communication scheme has to be implemented to achieve a high performance with that many CPUs.

In order to simulate the recrystallization of a large 3D microstructure, it is necessary to identify if and how the Pace3D solver scales. It is important to know the limits not only to determine the domain size or the subject complexity, but also to save resources like time, energy and money.

5.1. Phase-Field Implementation

Main components of the parallel Pace3D solver which are of interest for the performance analysis are explained in this section. The phase-field model presented by Nestler, Garcke and Stinner in [80], is the mathematical and physical basis for Pace3D and our investigations. In addition, phase transitions coupled with fluid dynamics, magnetic fields, elastic, plastic or volume preserved models are implemented in Pace3D, but are not considered in the following investigation. The implementation details give an overview of the relevant algorithms for the performance analysis.

The main workload of Pace3D is processed by the phase-field solver and its modules. The parallelized solver uses OpenMPI, an open source implementation of the common MPI standard. Updates of the fields are calculated with the explicit Euler scheme on a regular finite difference grid, such that 5 point stencil operations in 2D and 7 point stencil operations in 3D simulation domains, as shown in figure 5.1, are applied. The cuboidal simulation domain is decomposed for parallel computation, by dividing it into nearly equal-sized subdomains in the 1D case, as shown in figure 5.2, depending on the number of CPUs used.

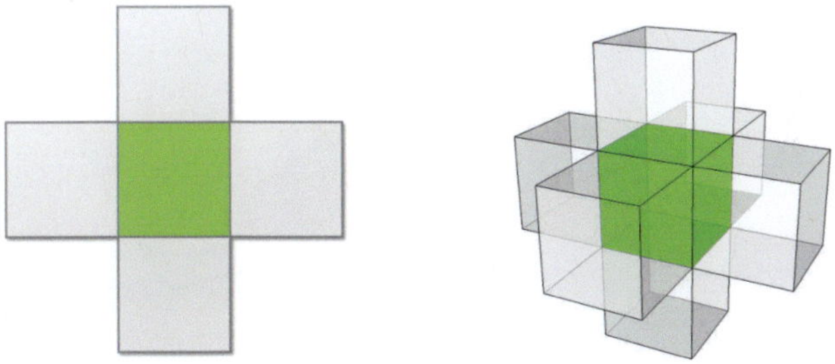

(a) 5 point stencil **(b)** 7 point stencil

Figure 5.1.: Computation stencils for (a) 2D and (b) 3D domains. The green cell in the center is updated by a computation, which includes the neighboring cells.

Ghost cells contain the field values of the neighbor subdomain to enable the stencil operations and consist of a line of cells in 2D and a plane of cells in 3D. Each subdomain can be processed by one CPU for one time step independently. After each time step, the cells at the cutting plane are transferred to update the ghost cells.

In order to control and administer the simulation, the Master-Worker pattern [63], as illustrated in figure 5.3, is implemented, so that each subdomain is assigned to one worker task. Each worker task is mapped to one CPU. The master task is responsible for initializing, controlling and stopping the worker tasks. Snapshots during a simulation are stored by the master in the 1D decomposition case, by receiving the subdomains of all worker tasks.

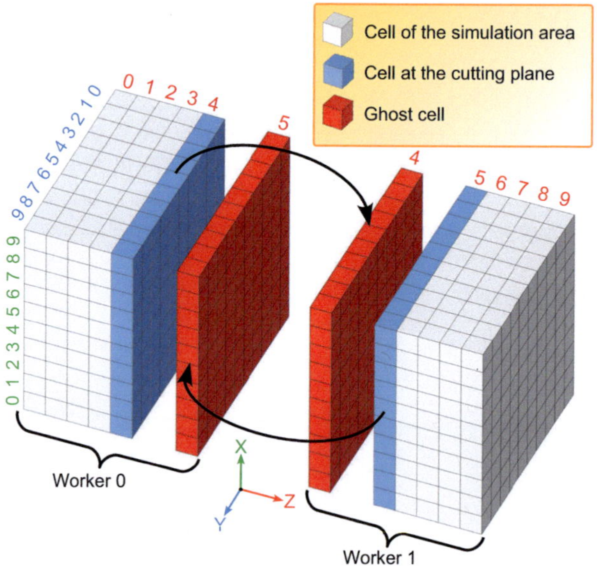

Figure 5.2.: 1D decomposition of a 10×10×10 simulation domain. The domain is divided into 2 subdomains and each subdomain is assigned to one CPU. The red colored ghost cells have to be updated by the neighboring CPU after each time step to enable the 16 point stencil operation inside each subdomain.

The ghost cell update is performed with blocking send and receive operations. It is a loose synchronization barrier, which involves every worker. Fast workers with a smaller workload are waiting for the ghost cell update of slower workers with a bigger workload. This imbalance is caused by the implementation of phase-field update calculation. Pure grain coarsening simulations for example do not contain a bulk entropy density term in the entropy functional, such that only the phase-field updates have to be computed. The phase-field equations are solved only in interface areas between phases and not in the bulk area. As a result, not every cell has to be computed and an imbalance of the workload in the simulation domain leads to CPU idle time of fast workers with a small workload.

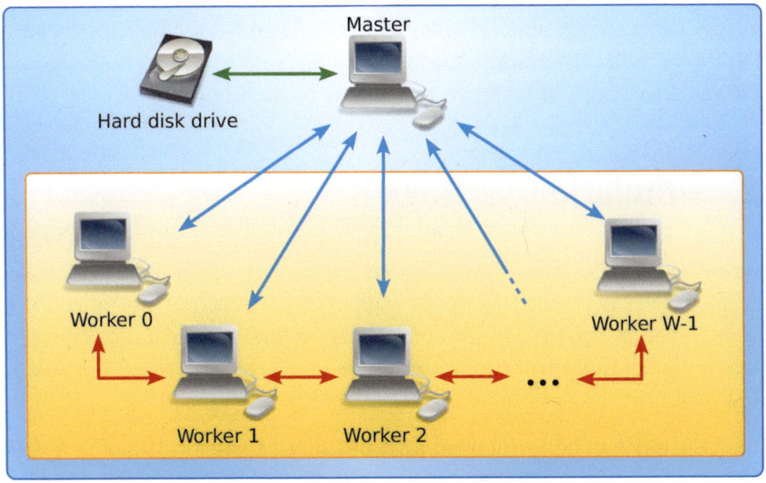

Figure 5.3.: Master-Worker pattern to organize and control the parallel simulation. The master controls all workers and stores simulation results on the hard disk. Between time steps, the workers exchange the boundaries directly without an intermediate master.

To reduce the CPU idle time, the load balancing mechanism dynamic domain decomposition (ddd) exists. Every worker measures its calculation time for a defined number of time steps and sends the accumulated calculation time to the master task. Based on the calculation time per worker, the master decides the optimal distribution of the domain and chooses which worker has to enlarge or reduce the subdomain. Figure 5.4 illustrates the 1D load balancing mechanism and the ddd procedure.

Another commonly used method to decrease computation time can only be applied if the area of interest moves along a trajectory during the simulation. Directional eutectic growth [103] in a temperature gradient, for example, is a solidification process which moves from one side of the simulation domain to the other. Crack propagation [108] is another example for a moving area of interest. A big simulation domain to capture the complete crack or the solidified area can be realized, however, it is not effective because a big part of the simulation domain has a small contribution but increases the simulation time and memory consumption. To avoid the increase of time and memory, a smaller simulation domain covering a larger one around the evolving phase boundary can be chosen, which moves with the solidification front (figure 5.5) or the crack tip. This tracking of the area of interest is referred to as moving box.

5.2. Simulation settings

Different applications are better suited for certain optimization techniques. Therefore, it is necessary to define different microstructure morphologies such as dendrite growth, grain coarsening and solidification of ternary eutectics in 3D to measure the performance of the implemented optimization techniques.

If information on the sequential and parallel execution times is gathered to obtain a performance description, the system specification (operating system, hardware properties, etc.) is necessary in order to make a well-founded comparison. The first system belongs to the Institute of Materials and Processes (IMP) of the University of Applied Sciences of Karlsruhe and the second belongs to the Scientific Computing Center of the Karlsruhe Institute of Technology (KIT). Hardware and software specifications are listed in table 5.1 and point out the main properties.

To evaluate the computational efficiency of the different optimization algorithms and strategies, a set of three typical microstructure simulations is defined with different requirements and characteristics for computational treatment. The considered cases are labeled with A, B and C in order to precisely refer to the simulation setup in the subsequent sections.

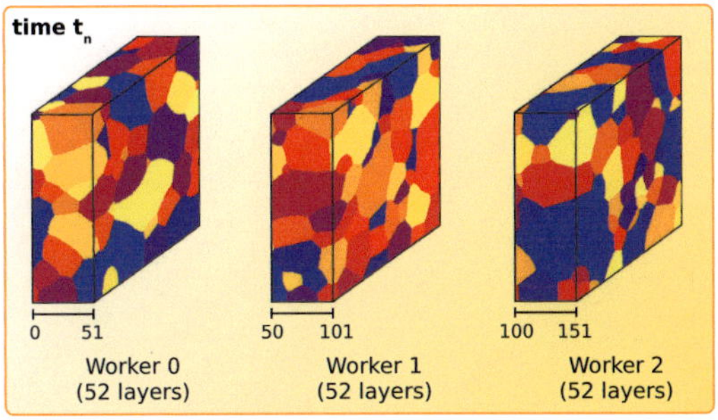

(a) Equally decomposed simulation domain with imbalanced load.

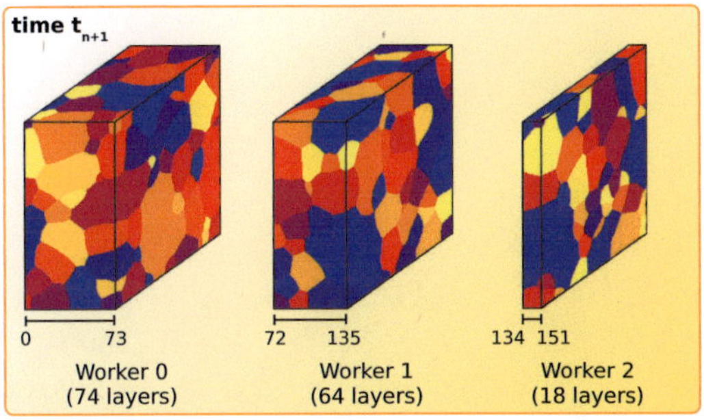

(b) Redistributed simulation domain with equally distributed load.

Figure 5.4.: Subdomain size before and after load balancing. (a) At time t_n, all workers have the same subdomain size (52 layers of XY planes). A high workload in the subdomain of worker 2 leads to an imbalance. (b) The domain of worker 2 is distributed to worker 0 and worker 1. Worker 0 is the fastest worker and gets more layers than the slower worker 1.

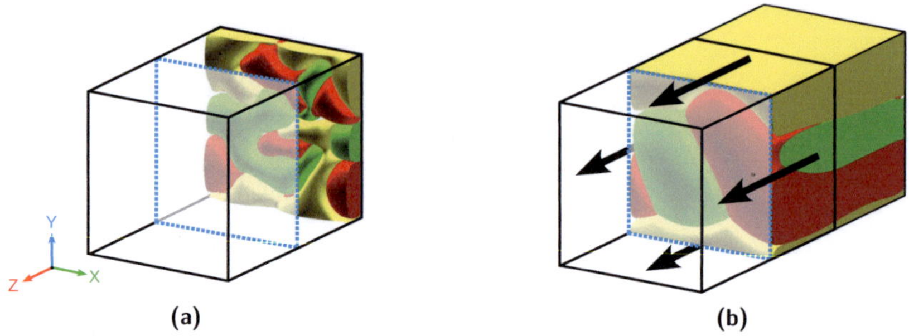

Figure 5.5.: A ternary eutectic at the rear side of the simulation domain (a) solidifies in z-direction. The simulation domain is shifted (b), if the solidification front reaches the trigger plane.

Hardware	IMP Cluster	XC 4000 Cluster
Compute nodes	48	750
	8 Cores	4 Cores
Cores per node	2 AMD Opterons (Quad-Core)	2 AMD Opterons (Dual-Core)8
Total cores	384	3000
Clock cycle	2 GHz	2.6 GHz
Main memory per core	2 GByte	4 GByte
Node interconnect	4X DDR Infiniband	4X DDR Infiniband
Interconnect Bandwidth	16 GBit/s	16 GBit/s
Interconnect Topology	Fat tree	Fat tree
Software		
Operation System	Debian Linux	Red Hat Linux
Compiler	GCC 4.3.2	Intel Compiler 10.1
MPI	OpenMPI 1.3.4	OpenMPI 1.5.1

Table 5.1.: Hardware and software specification of the utilized systems.

A Dendrite morphology

During the evolution of equiaxed dendritic structures, a complex surface geometry with the main trunk, the tip properties and secondary and ternary side branches is formed. Two dendrite simulation environments of a 2D and 3D domain are configured.

- A.1 2D domain of 1000×1000 cells with one nucleus in the center
- A.2 3D domain of 200×200×200 cells with one nucleus in the corner
- A.3 3D domain of 500×500×500 cells with one nucleus in the corner

In A.1, a 2D simulation of a single Ni-rich dendrite in a Ni-Cu melt with solutial solidification is set up, as done by Warren in [122], to compare the runtime between the IMP cluster and the XC 4000 cluster.

The same Ni-Cu system is used in A.2 for a single 3D dendrite to compare the parallel scalability of 1D and 3D domain decomposition due to small variations in computation time per cell. The evolution equation for the concentration fields are solved in each cell and take more computation time, in contrast to solving the phase-field evolution equations, which are only computed in the interface region. A.3 is only computed with 1D domain decomposition to compare 1D domain decomposition between A.3 and the 2D domain A.1.

A small domain of 200^3 cells is used to obtain the sequential runtime within a reasonable time frame. Larger simulation domains can be simulated, however, the runtime increases up to months for the sequential computation. To reduce the computational effort further, the use of the dendrite symmetry and placement of the nucleus at one corner of the domain is used. Figure 5.6 (a) shows the resulting 3D dendrite after mirroring along the symmetry planes.

B Directional solidification of a ternary eutectic

In directional solidification, most of the effects such as diffusion and phase transformations take place in regions around the solid-liquid front. Therefore, a high grid resolution is needed at the solidification front, combined with an increased numerical effort. A ternary eutectic solidification of three distinct solid phases (model system) grows in a dedicated direction, as depicted in figure 5.6b. Four ternary eutectic growth simulations are configured.

B.1 3D domain of 300×100×100 cells

B.2 3D domain of 100×100×300 cells

B.3 3D domain of 100×100×300 cells with load balancing

B.4 3D domain of 100×100×100 cells with moving box

Each configuration is executed with 100 workers to obtain comparable runtime performance. The differences of the four configurations are the domain orientation, domain size and the activated optimizations, such that the resulting eutectics are always equal, except for the moving box configuration. Dynamics outside the moving box is not captured, such that the results are not exactly equal.

By decomposing the domain along the z-axis for parallel computation, the domain orientation changes the amount of ghost cells. The influence of the communication time on the total runtime is determined. In the worst case (B.1) 300×100 cells and in the best cases (B.2-4) 100×100 cells have to be updated after each time step, respectively.

C Grain growth in a polycrystalline system

For grain coarsening, the bulk terms in the phase-field functional are neglected. While coarsening proceeds, the grain boundaries reduce, which leads to an uneven load in the computing cluster. It is an example suitable to measure the effect of dynamic domain decomposition. The following two grain coarsening domains are configured:

C.1 3D domain of 500×500×500 cells

C.2 3D domain of 250×500×1000 cells

The starting condition for C.1 is generated randomly with the Voronoi algorithm, as depicted in figure 5.6c. 125 workers calculate both configurations with and without load balancing. In contrast to the dendrite simulations, the maximum runtime is set to three days.

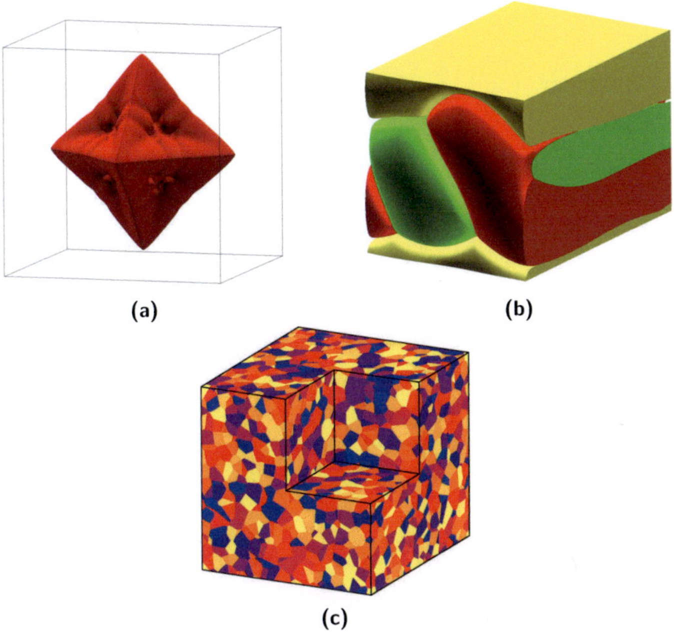

Figure 5.6.: Illustration of the three simulation cases: (a) Reconstructed 3D dendrite of configuration A.2 (b) Moving box of configuration B.4 (c) Voronoi tesselation generated according to the starting condition for C.1

5.3. Results

The measurement results for scalability, performance and dynamic load balancing of the sample settings A, B and C are outlined in this section.

A Scalability

Strong scaling behavior of A.1, A.2 and A.3 with 1D domain decomposition is depicted by the efficiency curves in figure 5.7. A list with all runtimes is provided in the appendix.

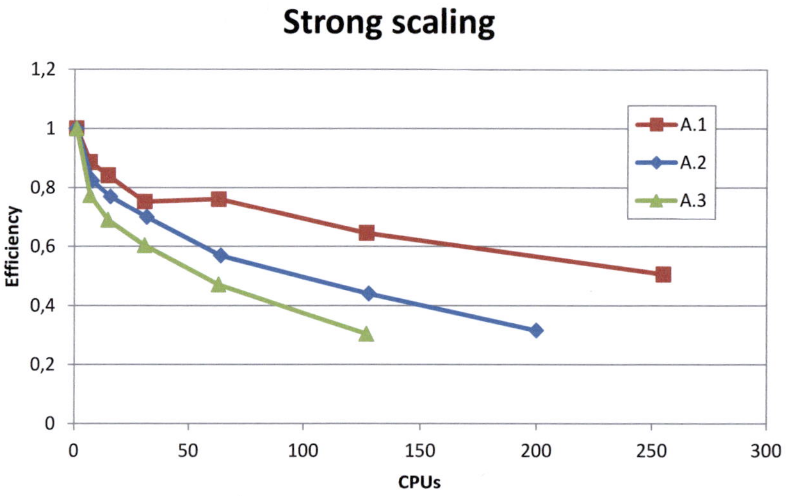

Figure 5.7.: Measured strong scaling behavior of A.1, A.2 and A.3. (a) 1D and 3D domain decomposition of A.2. (b) 1D domain decomposition of A.1 and A.3. Dashed lines denote the estimated scaling behavior with 3D domain decomposition.

The sequential and parallel 2D dendrite simulations A.1 are carried out on the IMP and XC 4 000 (short XC) cluster. Table 5.2 lists the runtime data for both clusters. On average, the simulations on the IMP cluster last 28,4% longer.

Workers (CPUs)	IMP	XC	IMP/XC
1 (1)	70:49:57	58:43:05	120,6%
8 (9)	11:25:16	09:30:17	120,2%
16 (17)	05:36:37	04:31:25	124,0%
32 (33)	03:02:31	02:16:06	134,1%
64 (65)	01:28:44	01:11:17	124,5%
128 (129)	00:51:56	00:39:05	132,9%
256 (257)	00:32:59	00:23:09	142,5%

Table 5.2.: Runtime of the 1 000×1 000 cells 2D dendrite simulation A.1 on the IMP and XC 4 000 cluster with varying CPU count and 1D decomposition.

B Communication bound simulation

If a non-cubic domain is used, the performance depends on the choice of the domain orientation. The ternary eutectics simulation B.1 has a runtime of 240 760 seconds. The rotated domain B.2 has a runtime of 142 731 seconds, which is 59.28% of the non-rotated domain B.1. With dynamic domain decomposition after every 100th time step (configuration B.3), the runtime decreases further to 129 394 seconds, which is 53.74% of B.1. Applying the moving domain algorithm to B.4, the runtime further decreases to 84 809 seconds, which is 35.23% of B.1. These timings are summarized in table 5.3 and indicate a communication bound character of the simulations.

C Load balancing

The runtime measurements of C.1 and C.2 with and without dynamic domain decomposition are shown in figure 5.8. The time to calculate 100 time steps is accumulated to one data point in the diagram. The elongated domain of C.2 seems to have a higher imbalance than the cubic domain.

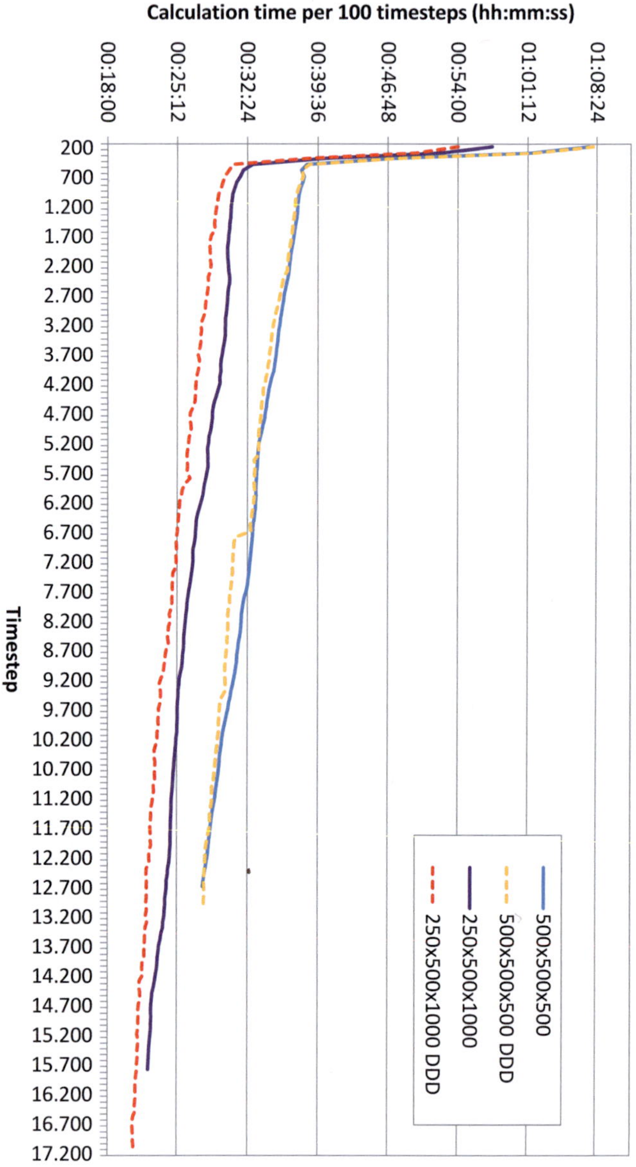

Figure 5.8.: Grain coarsening simulations C.1 and C.2 with a runtime limit of 3 days. Dashed lines denote the behavior with active load blancing.

Configuration	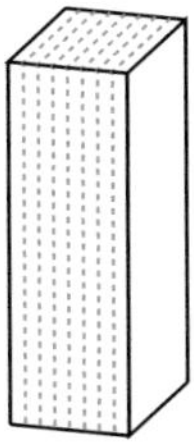	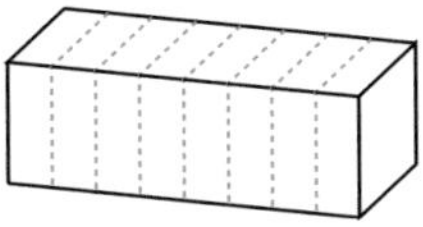
	B.1	B.2
Workers	100	100
Domain size	300×100×100	100×100×300
Optimization	-	-
Runtime d:hh:mm:ss	2:18:52:40	1:15:38:51
Runtime percentage	100%	59.28%

Configuration	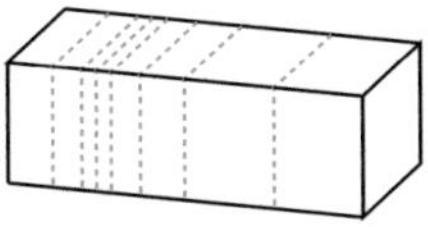	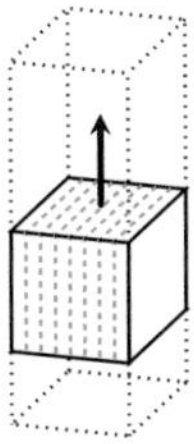
	B.3	B.4
Workers	100	100
Domain size	100×100×300	100×100×100
Optimization	Load balancing	Moving box
Runtime (d:hh:mm:ss)	1:11:56:34	0:23:33:29
Runtime (percentage)	53.74%	35.23%

Table 5.3.: Runtime of the 3D ternary eutectic configurations B.1-4.

5.4. Discussion

Strong scaling measurements in figure 5.7 illustrate the efficiency drop, which is related to the communication bound performance.

Comparing the runtime of the 2D dendrite simulations A.1 between the IMP and the XC cluster shows the advantage of the XC cluster due to the CPU frequency, because the clock cycle is the most outstanding difference of the systems. Considering the clock frequency, the CPUs of the IMP system need 30% more time than the CPUs of the XC cluster. Another issue is the configuration of the Infiniband network connecting the computing nodes and the file system, which could explain the performance gain of the XC cluster with many CPUs.

The influence of the domain orientation on the runtime is significant and can be observed within the ternary eutectic simulations B.1-4. The simulation of B.2 required only 59% of the simulation time of B.1. The main cause of the runtime differences can be related to communication, because the computation effort for B.1 and B.2 is the same. The boundary layer size is 100×100 for B.2 and 300×100 for B.1, respectively. In the latter case, the number of cells, which have to be transferred within the network, is three times larger for each computed time step of the iteration, which thus leads to at least three times longer communication time. Another reason for the speedup is a reduced amount of cache misses due to better memory alignment. Dynamic domain decomposition in B.3 further improves the runtime. The optimization of only calculating derivatives, if a gradient exists, seems to have an influence on the runtime, even with the long range concentration field. Comparing the moving box simulation B.4 and B.1, the runtime is almost 1/3, in accordance with the expectations. 1/3 less computation cells and 1/3 less communication lead to 1/3 of the runtime.

Load balancing shows, that the benefit strongly depends on the simulation content. Figure 5.8 indicates only one significant load imbalance for the C.1 simulation in a cubic domain ($500\times500\times500$) at about time step 6 800. The load distributes homogeneously after the imbalance, until about time step 13 000, which explains the convergence. The elongated domain C.2 ($250\times500\times1\,000$) has the advantage of moving smaller pieces of work, which enables a finer and hence better load balancing. All measurements contain a high computation time at the beginning, which is the result of the interface development during the first 100 time steps.

5.5. Conclusion

All measurements combined draw a rough performance picture of the implemented phase-field model with a clear direction for further improvements. The measurements indicate the communication bound character of the implementation, such that a reduction of the communication time is necessary to achieve a good strong scaling behavior on large high performance systems.

Beside the implementation details, it is of importance, which hardware is utilized for the simulation tasks. The investigations of the runtime on two similar clusters showed a different runtime behavior, which is mainly caused by the clock frequency of the CPUs.

1D load balancing reduces simulation time, but the benefit is strongly dependent on the simulation content and the degree of freedom. The load imbalance or the degree of freedom does not seem to be very high for the performed simulations. A major increase in load imbalance can be realized with the help of an adaptive mesh refinement scheme and with 3D domain decomposition. A proper investigation of load balancing strategies has to be performed to find general statements about the reachable benefit.

To obtain a deeper insight of the simulation runtime on high performance computers and to fine-tune the implementation, it will be necessary to apply elaborated profiling tools like VampirTrace [77] or Scalasca [27]. Further steps to increase the scalability are 3D domain decomposition, communication hiding, 3D load balancing and adaptive mesh refinement.

6. Optimization

The measurements of the performance analysis show that the phase-field implementation is communication bound. To optimize the implementation, it is necessary to reduce the amount of data, which has to be communicated. 3D domain decomposition, also known as 3D blocking [82], reduces the amount of data, which has to be transferred. This chapter describes the implemented 3D domain decomposition and the formulated performance model to estimate the scaling behavior for simulations with more than 10 000 CPUs. Measurements are provided to validate the optimization.

The phase-field community has discovered many optimizations to improve the performance, but a rigorous performance analysis of the scaling behavior could not be found, such that an own study has to be performed. It turned out, that decomposing a domain along one dimension for parallel execution in subdomains is a considerable bottleneck, because the surface to volume ratio is too high. An efficient execution of a 3D simulation with more than 500 CPUs is not possible.

To improve the performance, it is necessary to decompose the simulation domain along all three dimensions, which reduces the surface A to volume V ratio $\frac{A}{V}$ of the subdomains. The surface of a subdomain is directly related to the amount of data, which has to be communicated, because the cut surface has to be updated between subgrains. This can be seen in figure 5.2, where the domain is decomposed in one dimension along the z-axis, such that the cutting surface has to be updated each time step.
The surface to volume relation for solid objects is well understood and a famous example from biology is Bergmann's rule. In 1848, Bergmann described [11], that animals of the same family e. g. penguins in cold regions are bigger than in warm regions of the world, because the surface to volume ratio helps to save energy. The surface per volume or mass is smaller for bigger penguins such that less energy per volume has to be

used to hold the body temperature. A simplified calculation of the surface to volume ratio for the Humboldt penguin and the Emperor penguin is shown in table 6.1.

Humboldt Penguin	Emperor Penguin

r = 5 cm	r = 10 cm
h = 45 cm	h = 100 cm
A = 15.7 dm^2	A = 69.1 dm^2
V = 3.5 dm^3	V = 31.4 dm^3
$\frac{A}{V} = 4.4$	$\frac{A}{V} = 2.2$

Table 6.1.: The surface to volume ratio $\frac{A}{V}$ of the Emperor penguin helps to survive at the cold South Pole. The smaller Humboldt penguin has a bigger surface to volume ratio and lives in warmer South America.

The same relation holds for subdomains of a decomposed domain for parallel execution. A 3D decomposition of the domain reduces the surface to volume ratio of the subdomains compared to 1D decomposition. For parallel execution, it is sufficient to only consider one subdomain as a simplification, because each subdoamin is computed in parallel and should have nearly the same shape. The same calculation as for the penguins is provided in table 6.2 for a subdomain of a 1D and 3D decomposed domain with 8 subdomains ($W = 8$) and a cube size of 1.

3D domain decomposition reduces the amount of data, which has to be transferred from one CPU or worker (W) to the other not by changing the volume of the subdomains but by changing the surface. Figure 6.1 shows the 3D domain decomposition of a simulation domain in x-, y- and z-axis into 8 subdomains with ghost cells in red and cells which are transferred to update the ghost cells in blue.

1D decomposition	3D decomposition

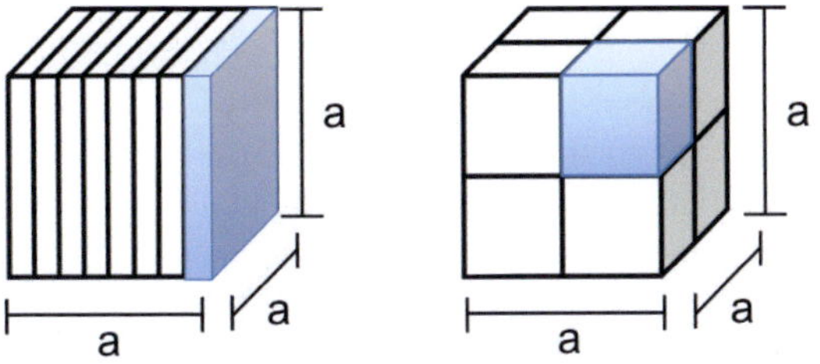

1D decomposition	3D decomposition
a = 1	a = 1
W = 8	W = 8
A = 2.5	A = 1.5
V = 0.125	V = 0.125
$\frac{A}{V} = 20$	$\frac{A}{V} = 12$

Table 6.2.: A cubic simulation domain is decomposed into 8 subdomains with 1D (left) and 3D (right) domain decomposition. The surface to volume ratio for the blue subdomains is bigger for 1D decomposition than for 3D decomposition.

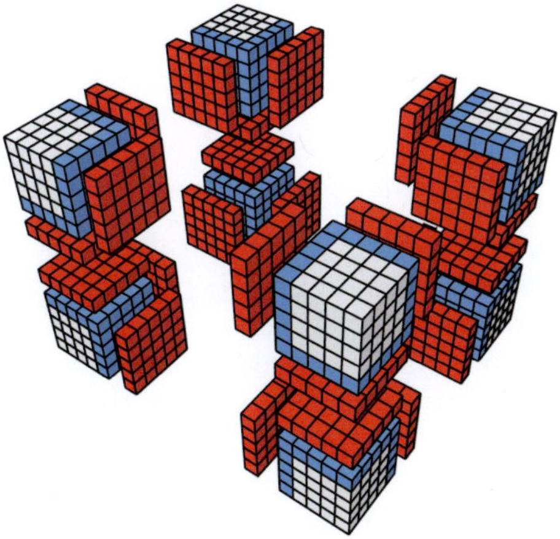

Figure 6.1.: 3D domain decomposition also known as 3D blocking of a 10×10×10 simulation domain. Ghost cells are red and cells which are transferred to update ghost cells are blue.

6.1. Performance modeling

A performance model enables estimations of the scaling behavior without the need to perform a complete simulation series. To estimate the runtime, it is assumed, that the size of the simulation domain, the computation time and the communication time have a significant influence on the runtime. The runtime is divided into calculation, communication and idle time, which are dependent on the subdomain size of the workers. Equation (6.1) describes the simulation time G^w for a worker w and over all time steps τ_{max} with the three terms related to computation, communication and idle time,

$$G^w = \sum_{\tau=1}^{\tau_{max}} \overbrace{n_c^\tau \cdot O^\tau}^{\text{Computation}} + \overbrace{\left(n_g^\tau + n_l^\tau\right) \cdot T^\tau}^{\text{Communication}} + \overbrace{I^\tau}^{\text{Idle}} . \tag{6.1}$$

All parameters depend on the time step τ in the general case. The computation time is estimated by the number of cells n_c^τ, a worker contains in its subdomain, times the average computation time per cell O^τ in the subdomain. To model the communication time after each time step, it is assumed that all ghost cells n_g^τ of the neigboring subdomain are updated with the transfer time T^τ per cell. Communication time for load balancing is integrated by the number of cells n_l^τ, a domain is enlarged and reduced. Blocking communication together with load imbalance involves idle time I^τ, which is taken into account by the last term.

This model allows the analysis of the parallel scaling behavior incorporating the knowledge of the time-dependent functions, which is not always given because of the dynamic simulation content. To estimate the effect of different domain decomposition methods, the model is simplified and it is assumed, that all parameters are independent of the time step τ and that a cubic simulation domain of $n \times n \times n$ cells is used; each worker has the same subdomain size, load balancing is not considered and each cell in the simulation domain has the same computation and communication time, such that idle time can be neglected.

The sequential runtime G_{seq} is described on a single CPU by the total amount of cells, by the calculation time per cell O and by the number of time steps τ_{max}, according to the expression

$$G_{seq} = n_c \cdot O \cdot \tau_{max}. \tag{6.2}$$

For parallel execution (without load balancing and idle time), the communication term $n_g \cdot T \cdot \tau_{max}$ is added, which reads as

$$G^{w}_{par} = (n_c \cdot O + n_g' \cdot T)\, \tau_{max}. \tag{6.3}$$

The ghost cell update between two neighbors is idealized to one send-receive operation, assuming an ideal full duplex connection between workers. In case of 1D domain decomposition, each worker has to update ghost cells of the left and right subdomain, such that $n_g = 2 \cdot n^2$. For 3D domain decomposition, each worker has to update a fraction of the ghost cells for 1D domain decomposition and to update the 6 neighboring ghost cell areas, such that $n_g = 6 \cdot \frac{n^2}{\sqrt[3]{W}^2}$. The runtime for 2D domain decomposition is created according to the 1D and 3D case, where 4 neighbors have to be updated with $n_g = 4 \cdot \frac{n^2}{\sqrt{W}}$. The domain size is constrained, so that it is equally divided by the number of workers W without a remainder. Relying on these assumptions, each worker has the same runtime. To predict the scaling behavior, it is only necessary to model the runtime of one worker. Finally, the runtime model of a worker with 1D, 2D and 3D domain decomposition is

$$G_{1D} = \left(\frac{n^3}{W} \cdot O + 2 \cdot n^2 \cdot T\right) \tau_{max} \tag{6.4}$$

$$G_{2D} = \left(\frac{n^3}{W} \cdot O + 4 \cdot \frac{n^2}{\sqrt{W}} \cdot T\right) \tau_{max} \tag{6.5}$$

$$G_{3D} = \left(\frac{n^3}{W} \cdot O + 6 \cdot \frac{n^2}{\sqrt[3]{W}^2} \cdot T\right) \tau_{max}. \tag{6.6}$$

To avoid invalid configurations, the domain size and the number of workers is restricted, such that the results of $\frac{n^3}{W}$, $\sqrt[3]{W}$ and $\sqrt{W}$ are integer. This restriction limits the domain extensions and the number of workers, but it allows an easier investigation of the parallel scaling behavior.

To estimate the scaling behavior of the different dimensions with 1D, 2D and 3D domain decomposition, the generalized formulation is derived from the simulation runtime for a Ω-dimensional domain with a d-dimensional domain decomposition by

$$G_d^\Omega = \left(\frac{n^\Omega}{W} \cdot O + 2 \cdot d \left(\frac{n}{\sqrt[d]{W}} \right)^{d-1} \cdot n^{\Omega - d} \cdot T \right) \tau_{max} \tag{6.7}$$

with the restriction $\Omega \geq d$.

Figure 6.2 compares the 1D and 3D decomposition methods with 64 workers operating on a cubic domain of $n \times n \times n$ cells to explain the origin of the restrictions and terms of the models. Figure 6.2 (c) compares the amount of cutting plane cells in higher order direction to illustrate the communication advantage of the 3D decomposition method for 3D domains. Each worker has the same amount of cells to compute, but a different amount of boundary cells to transfer.

To compare the efficiency of the decomposition methods dependent on the worker count, the general function $E(W)_d^\Omega$ is derived based on equation (6.7) with the common definition of speedup ($S = \frac{G_{seq}}{G_{par}}$) and efficiency ($E = \frac{S}{W}$).

$$E(W)_d^\Omega = \frac{S}{W} = \frac{\frac{G_{seq}}{G_{par}}}{W} = \frac{G_{seq}}{G_{par} \cdot W} = \tag{6.8}$$

$$= \frac{n^\Omega \cdot O}{\left(\frac{n^\Omega}{W} \cdot O + 2 \cdot d \cdot \left(\frac{n}{\sqrt[d]{W}} \right)^{d-1} \cdot n^{\Omega - d} \cdot T \right) W} \tag{6.9}$$

$$= \frac{1}{1 + \frac{2d}{n} \cdot \frac{T}{O} \cdot \sqrt[d]{W}} \tag{6.10}$$

This model abstracts the underlying hardware, network and software properties to communication time per cell T and computation time per cell O. The simplification allows to determine the scaling behavior by the communication to computation time ratio $\frac{T}{O}$. Equation (6.8) is a root function, which is not dependent on the dimensionality of the domain and reveals the scaling behavior.

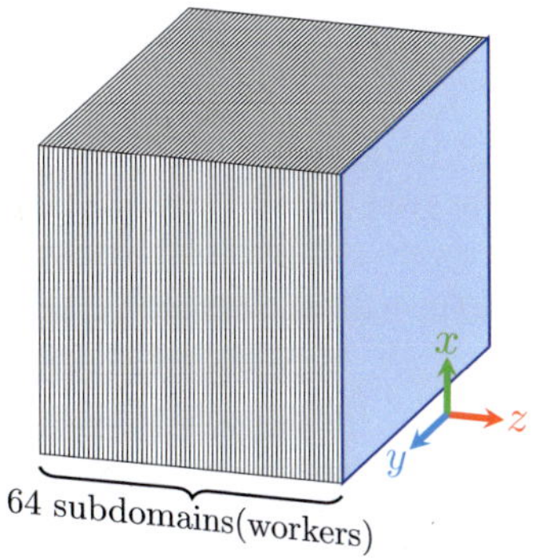

(a) 1D domain decomposition

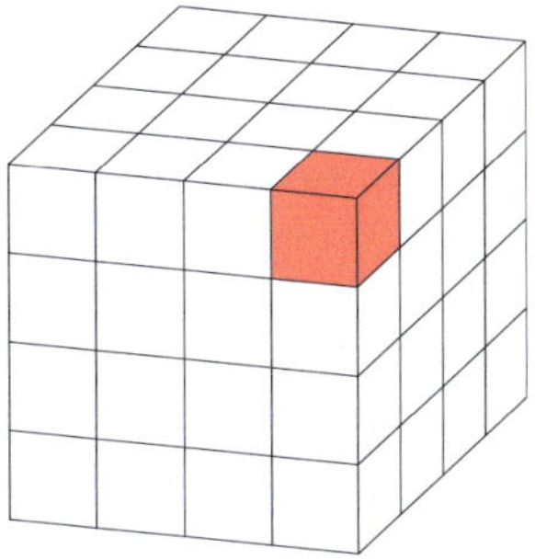

(b) 3D domain decomposition

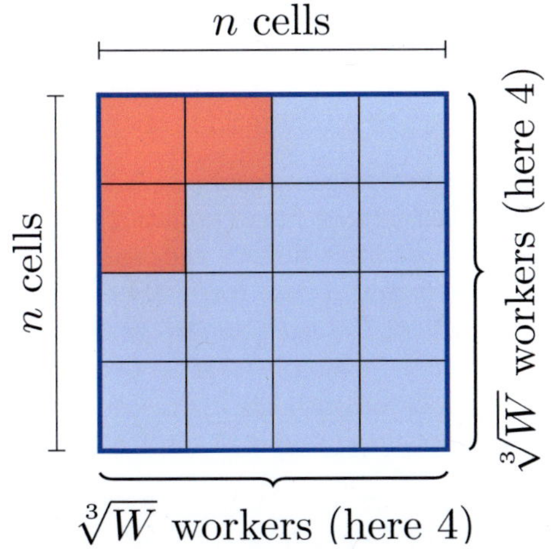

(c)

Figure 6.2.: 1D and 3D decomposition with ghost layers for a simulation domain of $n \times n \times n$ cells with 64 workers. (a) 1D decomposition along the z-axis with 64 workers to 64 subdomains. The blue area is one ghost surface. (b) 3D decomposition of the same domain with the same amount of workers and 3 red ghost surfaces. (c) The ghost surfaces of the 1D and 3D decomposition are related to each other. Only $\frac{3}{16}$ of the 1D decomposition surface has to be updated with 64 workers, in case of 3D domain decomposition.

The limit of equation (6.8), for an infinite number of workers, is always 0 but the question is: How fast is the efficiency dropping?

The efficiency drop is mainly determined by how fast W, $\sqrt{W}$ and $\sqrt[3]{W}$ are growing. The solution is, that $\sqrt[3]{W}$ is growing the slowest, hence the efficiency drops the slowest. This property, described by the performance model, is supported by performance measurements presented in the next section.

6.2. Measurement results

The simulation setup A.2 is computed in parallel with 3D domain decomposition to compare the influence of the domain decomposition on the runtime. Estimations for A.1 and A.3 are provided together with the estimated communication to computation ratios $\frac{T}{O}$.

Figure 6.3 depicts the measurement results of A.2 with 1D and 3D domain decomposition. The least square fits of equation (6.4) and equation (6.6) are provided. Thin black lines denote the least square fit of equation (6.8) with $\Omega = 3$, $d = 1$ and $d = 3$ for a 3D domain with 1D and 3D decomposition. A result of the least square fit is the communication to computation parameter $\frac{T}{O}$, which is used to predict the performance of the 3D domain decomposition with the 1D measurements. The parameter $\frac{T}{O}$ is 1.08 for 1D decomposition and 0.73 for 3D domain decomposition.

A deviation of the measurements from the least square fit in figure 6.3 and figure 6.4, for 1D decomposition with up to 50, is the result of a non-load balanced simulation. The performance model assumes an equal distribution of the load but the simulation of a growing nucleus in the corner of the domain keeps the computation load not in balance. This deviation from the model is bigger for a small amount of CPUs than for a large amount. The more CPUs are considered, the faster the load is decomposed across the CPUs.

These measurements reflect a small part of the scaling behavior. It is of interest to determine the overall scaling performance of the 1D and 3D decomposition methods for clusters with more than 100 000 CPUs. Based on the presented performance model, the efficiency surface is plotted over the number of CPUs and over the communication to computation coefficient $\frac{T}{O}$, as prescribed by the equation (6.7) for a 3D domain ($\Omega = 3$), with 1D ($d = 1$) and 3D ($d = 3$) domain decomposition. The aim of using big systems is the ability of simulating complex problems, to resolve phenomena with a higher resolution or to describe scale bridging effects. A cubic domain of $5\,000 \times 5\,000 \times 5\,000$ cells is assumed to estimate the scaling behavior of 1D and 3D domain decomposition in large computational domains and to plot the results in figure 6.5 and figure 6.6.

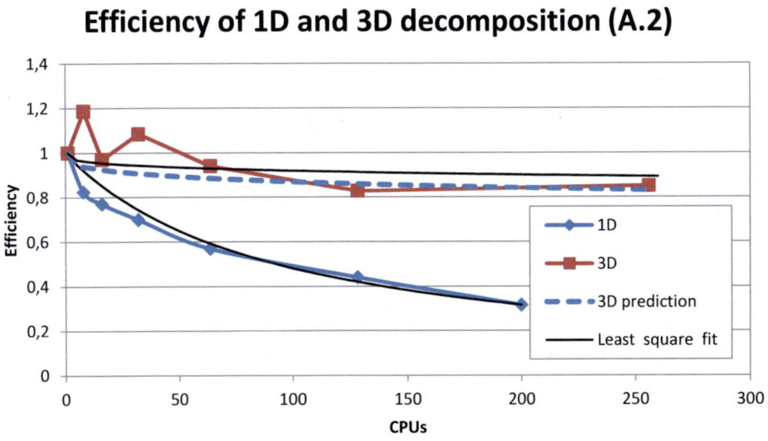

Figure 6.3.: Efficiency of A.2 with 1D and 3D domain decomposition. A prediction based on the communication to computation parameter $\frac{T}{O}$ of the 1D decomposition measurements is denoted by a dashed line.

These predictions are only valid if the communication to computation parameters $\frac{T}{O}$ are big enough. If the network is very fast, then T is small and it makes almost no difference which decomposition method is used, beside the amount of usable processors. Table 6.3 contains the 5 calculated coefficients based on the measurements. B.1 and B.2 have the same $\frac{T}{O}$ because a set of linear equations is solved, which is derived from equation (6.4). The derived equation for a non-cubic domain and 1D decomposition along the z-axis is

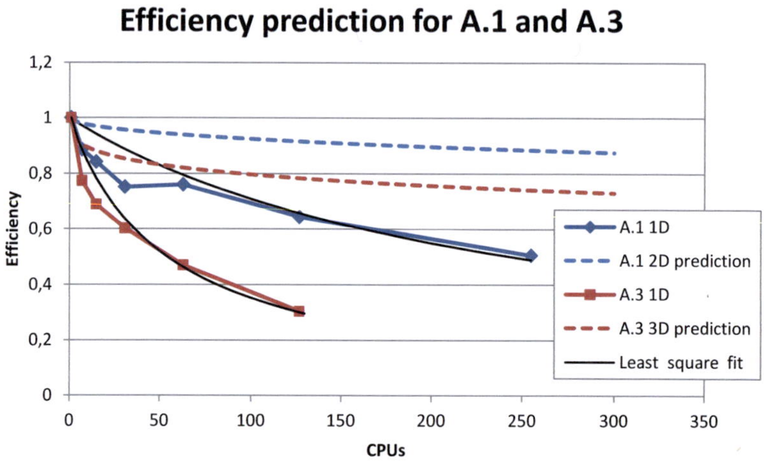

Figure 6.4.: Efficiency of A.1 and A.3 with 1D and 3D domain decomposition. A prediction based on the communication to computation parameter $\frac{T}{O}$ of the 1D decomposition measurements is denoted by a dashed line.

$$G_{1D} = \left(\frac{n^3}{W} \cdot O + 2 \cdot n^2 \cdot T \right) \tau_{max} \tag{6.11}$$

$$G_{1D}^{\mathrm{B.X}} = \left(\frac{n_x \cdot n_y \cdot n_z}{W} \cdot O + 2 \cdot n_x \cdot n_y \cdot T \right) \tau_{max}. \tag{6.12}$$

The set of linear equations based on the measurements B.1 and B.2 in table 5.3 is

$$\mathrm{B.1} : 240\,760 = \left(\frac{300 \cdot 100 \cdot 100}{100} \cdot O + 2 \cdot 300 \cdot 100 \cdot T \right) \tau_{max} \tag{6.13}$$

$$\mathrm{B.2} : 142\,731 = \left(\frac{100 \cdot 100 \cdot 300}{100} \cdot O + 2 \cdot 100 \cdot 100 \cdot T \right) \tau_{max}, \tag{6.14}$$

such that $\frac{T}{O}$ is 0.78 for B.1 and B.2.

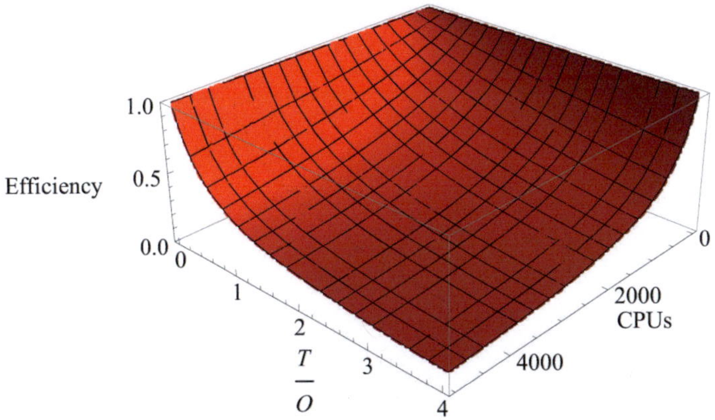

Figure 6.5.: Efficiency surface of 1D domain decomposition with a domain of 5 000×5 000×5 000 cells. 1D decomposition limits the amount of usable CPUs to 5 000 and drops fast to a low efficiency.

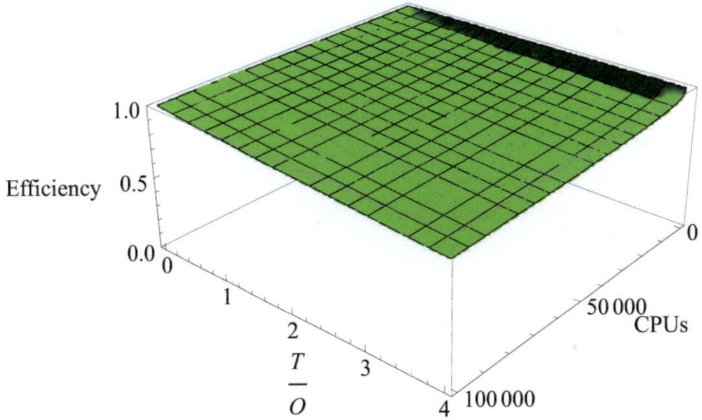

Figure 6.6.: Efficiency surface of 3D domain decomposition with a domain of 5 000×5 000×5 000 cells. A high efficiency is maintained up to 100 000 CPUs.

Setup	1D decomposition	3D decomposition
A.1	2.05	-
A.2	1.08	0.63
A.3	4.59	-
B.1 & B.2	0.78	-

Table 6.3.: Estimated and calculated communication to computation coefficients $\frac{T}{O}$ for 1D and 3D domain decomposition (short 1DDD and 3DDD).

6.3. Discussion

The efficiency drop with 1D domain decomposition occurs faster for 3D domains than for 2D domains, as shown in figure 6.4. This is a result of the dimensionality of the decomposition in relation to the dimensionality of the domain. The higher the difference, the earlier the efficiency drop happens. Considering a domain with 1 million cells in 1D, 2D (1 000×1 000) and 3D (100×100×100), the amount of transfer cells per worker is 1 for 1D, 1 000 for 2D and 10 000 for the 3D domain with 1D domain decomposition. Figure 6.7 shows this issue with the performance model and a $\frac{T}{O}$ of 1.

Figure 6.8 depicts the improvement of the efficiency for a 3D domain, which changes the domain decomposition method. The smaller the volume to surface ratio, the better the scaling behavior. Figure 6.5 and figure 6.6 show the significantly different scaling behavior between 1D and 3D domain decomposition, as well as the limits of 1D domain decomposition for a 3D domain.

Superlinear speedup is measured for A.2 with 3D domain decomposition, which could be caused by cache effects or by an ideal mapping of the subdomains on the computation nodes. The superlinear speedup is not occurring or not pronounced for the simulation with 256 CPUs.

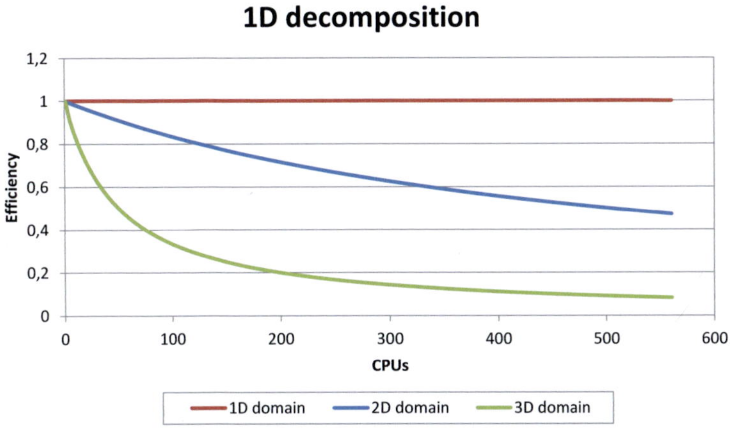

Figure 6.7.: Efficiency estimation with the performance model of 1D domain decomposition and a 1D, 2D and 3D domain. Each domain has 1 million cells, such that n=1 000 000 for the 1D, n=1 000 for the 2D and n=100 for the 3D domain. The communication to computation ratio $\frac{T}{O}$ is 1. The efficiency drops faster the higher the dimension of the domain.

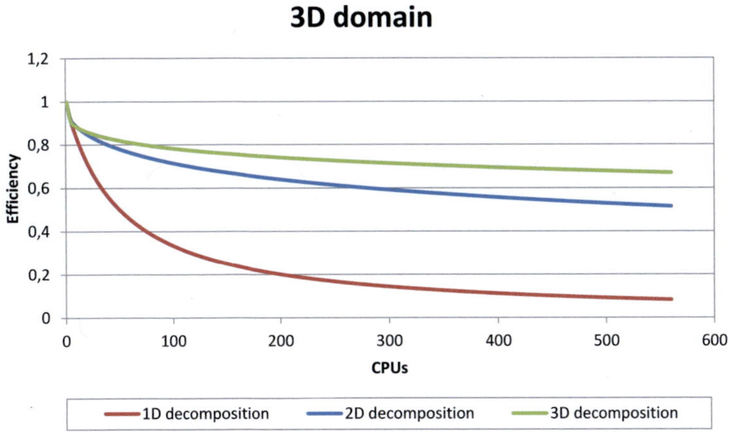

Figure 6.8.: Efficiency estimation for 1D, 2D and 3D domain decomposition of a 3D domain. The 3D decomposition has a better efficiency.

6.4. Conclusion

3D domain decomposition is a necessary milestone to perform large simulations with many CPUs because it increases the parallel performance significantly compared to 1D domain decomposition.

The measurements support the assumption, that communication is the limiting factor for the scaling behavior of the Pace3D solver. Figure 6.3 shows, that 3D domain decomposition can halve the simulation time at 128 CPUs and that predictions based on 1D domain decomposition match well with the measurements. The efficiency of 3D domain decomposition is always higher than for 1D domain decomposition and superlinear speedup can be achieved.

The performance model is well suited to estimate and to predict the scaling behavior without the necessity to model hardware specific properties. Estimations within a valid range of the communication to computation parameter $\frac{T}{O}$ from table 6.3 are shown in figure 6.5 and 6.6 to predict the scaling behavior for 5 000 and 100 000 CPUs with 1D and 3D domain decomposition of large domains. It pictures the advantages of the 3D domain decomposition over 1D domain decomposition. More CPUs can be utilized and the efficiency stays high.

7. Recrystallization

The numerical study of recrystallization on the microstructure scale in three dimensions is a challenging task. The stored energy introduced during deformation by dislocations has to be estimated well in order to create the starting conditions for a subgrain-based simulation of recrystallization. The deformation process has to be considered, because the stored energy is strongly related to the recrystallization characteristics, such as subgrain-based nucleation rate, nucleation sites and growth speed. Moreover, preferred texture components develop with a specifically stored energy distribution during the deformation process, according to the properties of the grain structure. Unfortunately, no comprehensive model is able to describe both, large deformations and grain structure dynamics, as well as the current crystal plasticity and free surface models, so that models and methods have to be coupled in order to describe recrystallization with state-of-the-art tools.

Because of the interaction of three scientific fields, some contributions in this chapter are performed by others. Dr.-Ing. Simone Schreijäg and Dr. Reiner Mönig of the Karlsruhe Institute of Technology provided the experimental data described in the section 7.1.1. The rolling simulation described in section 7.1.2 with a crystal plasticity finite element model is provided by Pierre Bienger and Dr. Dirk Helm of the Fraunhofer IWM in Freiburg Germany.

The following sections give a brief overview of major developments in crystal plasticity theory and methods to simulate recrystallization. Then, a short summary of coupled approaches is given to round up the introduction.

The knowledge of microstructure, its crystallography and the knowledge of grain boundary motion have increased over the past 100 years. Johnson, Mehl, [6] and Kolmogorov developed the analytical JMAK model to study recrystallization. It takes the nucleation rate, dimension, growth speed and growth shape into account to describe a recrystallization volume fraction over time. The disadvantage is that the underlying microstructure is not considered. Numerical Monte Carlo models, like the Potts model and the large Q-Potts model by Potts [87], are able to incorporate the preexisting microstructure and a stored energy field [109] to describe recrystallization in two and three dimensions [41, 75]. Cellular automata fall into the same category and are a fast, rule-based method to study crystal evolution and recrystallization like von Neumann [119] and Raabe [89]. In contrast to the methods with regular grids, vertex models by Weygand et al. [127] explicitly represent grain boundaries with curves and surfaces to investigate microstructure phenomena. One of the youngest methods to study recrystallization is the phase-field method.

Regardless of the method that is used to simulate recrystallization, the effort required to measure and thus obtain the polycrystalline grain structure with the stored energy for 3D simulations is high. Coupled approaches of crystal plasticity models with free boundary models are able to contribute to closing this gap. Despite the interests in using one discretization method, as is done by e. g. Li et al. [62] or Takaki et al. [111], the focus of the following summary is placed on coupled discretization methods to incorporate or study recrystallization.

The coupling of a crystal plasticity finite element model with a cellular automaton to simulate static recrystallization is presented by Raabe and Becker [90] in two dimensions and is discussed for three dimensions in the publication of 2002. A measured grain structure is used as the starting condition for the deformation and a Voronoi tessellation is utilized to couple the discretization methods after the deformation. The stored energy and the crystal lattice orientations of each finite element integration point are mapped on the regular cellular automaton grid. The accumulated shear is linearly related to the stored energy and the dislocation density, respectively, to estimate the driving force for nucleation and growth. The study of recrystallization for single and bicrystals of aluminum by Radhakrishnan and Sarma in 2000 and 2004, is performed in the same manner by coupling crystal plasticity with a Monte Carlo method. Radhakrishnan and Sarma use the slip system shear hardening values to estimate the stored energy, as Raabe and Becker.

Muramatsu et al. [79] present a two-dimensional dislocation-based approach to capture and to incorporate the subgrain structure. Their study is based on the crystal plasticity model with dislocation diffusion by Yamaki et al. [130]. 2D phase-field simulations are conducted to simulate nucleation and growth based on the present dislocation density distribution. Another 2D phase-field related approach which incorporates subgrains is presented by Takaki and Tomita [112]. The subgrain structure generation is dependent on the stored energy and the evolution of subgrains is computed without an additional driving force, such that no explicit nuclei have to be determined. The idea of the abnormal grain growth approach for nucleation is analyzed by Humphreys [44] and used by Suwa et al. [110] to simulate recrystallization without a stored energy.

Recent coupled approaches show refinements that make it possible to map the stored energy more accurately [34] by interpolation, to incorporate 3D simulations [55] and to gain insights in material behavior. The coupled approach in the detailed study by Won Lee and Im [128] is a successful application to investigate material behavior during processing with dynamic recrystallization by cellular automata in 2D.

In order to contribute to the development, a coupled approach of a crystal plasticity finite element model and the phase-field method to simulate cold rolling (deformation) with subsequent annealing (static recrystallization) at the end of the sheet metal production chain in 3D incorporating 2D electron backscatter diffraction (EBSD) measurements is presented. The emphasis lies in the development of the orientation distributions with large-scale simulations to provide more reliable orientation distributions with a coupled approach.

7.1. Method

The following three methodological paths are distinguished: experiments, finite element method and phase-field method, as shown in figure 7.1[5]. The experimental measurements are used to generate starting conditions, to parameterize the simulations and to compare the simulation results. EBSD

[5] The figures in this chapter are taken from the submitted manuscript in Computational Mechanics (Springer) journal entitled .

measurements and tensile tests after each production step make up the data basis. The cold rolling simulation is performed with a crystal plasticity finite element model. EBSD data is used to create the initial microstructure in 3D. In this initialization procedure, the crystal orientations are mapped to a 3D domain in a statistically equivalent manner, which is tessellated by the Voronoi algorithm. Cold rolling simulations and subsequent tensile tests have been carried out in order to identify the hardening parameters by the inverse simulation technique, where the hardening parameters are varied until an adequate agreement between the experimental and simulated stress-strain curve after cold rolling is obtained. The cold rolling simulation results are compared with the crystallographic texture information of the EBSD measurement. To simulate recrystallization, the crystal lattice orientations and the stored energy of the finite element simulation are mapped to the regular finite differences grid, as described by Raabe and Becker [90]. The final states of recrystallization are compared with the texture information of the EBSD measurements.

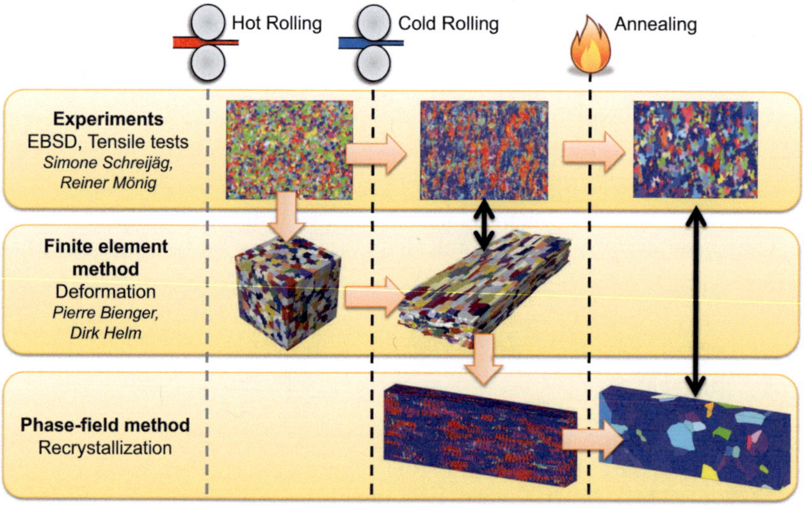

Figure 7.1.: Overview of the experimental and numerical process chain showing the data flow and the incorporated simulation methods. The simulation methods, together with the experiments, are ordered vertically in layers. The processing steps rolling and annealing are ordered horizontally.

7.1.1. Experimental investigation

The experimental investigations are performed with non-alloyed DC04 steel, which is a standard deep drawing steel in the automotive industry. DC04 can withstand high fracture strains of up to 50% and yields at stresses between 300 MPa to 350 MPa. During the cold rolling process, the thickness of the sheet plate is reduced by 62% from 3.2 mm to 1.2 mm. Besides low impurity concentrations, DC04 contains 0.08wt% C and 0.40wt% Mn, as described by Wegst and Wegst in [124]. The material mainly consists of a ferrite (bcc) phase containing Fe and low contents of C. Very small α-MnS precipitates with diameters below 200 nm and spacings of a few micrometers are found inside the ferrite grains.

To determine the microstructure of the material after each process step, EBSD measurements are performed. In order to obtain sample surfaces that are sufficiently smooth, the DC04 specimens are mechanically ground using SiC paper of decreasing particle size and polished with 6 μm, 3 μm and 1 μm diamond paste. In order to remove the remaining deformations that are introduced by the mechanical polishing, the samples are polished electrochemically.

7.1.2. Crystal plasticity finite element model

In the crystal plasticity model based on Asaro [2, 3], the occurring energy storage phenomena (cf. e. g. [36]), given by elasticity and dislocation-based plasticity, are not taken into account in detail. However, for the modeling of the static recrystallization in section 7.1.4, at least the stored plastic energy must be estimated in a useful way, because the stored energy is the most important driving force for the primary recrystallization processes: From metal physics, the energy storage phenomena are well understood, due to the production and movement of dislocations, and are usually formulated in terms of the dislocation density (cf. e. g. Dejan Stojakovic: Microstructure Evolution in Deformed and Recrystallized Electrical Steel, Dissertation, Drexel University, 2008). In metal physics, the critical resolved shear stress τ is proportional to the dislocation density ρ:

$$\tau \sim \sqrt{\rho}. \tag{7.1}$$

Using the dislocation density, the contribution to the stored energy by dislocations can roughly be estimated with $E_{\text{store}} \sim \rho \sim \tau^2 \sim \Gamma^2$ by using the hardening function, which depends on the cumulative shear Γ. As a result, the ansatz is reached to roughly approximate the stored energy with the cumulative shear by summarizing all constant values in $\tilde{E}_0$ so as to obtain

$$E_{\text{store}} = \tilde{E}_0 \Gamma^2. \tag{7.2}$$

Consequently, the discussion results in the well-known fact that, in a phenomenological sense, the stored energy can be considered as a function of the accumulated plastic strain. Therefore, only the orientation and the internal state variable Γ^2, as the stored energy, are transferred to the phase-field method for recrystallization.

The contribution of residual stresses to the stored energy is neglected because the dominant contribution to the stored energy for recrystallization comes from the dislocations. The residual stresses reduce during polygonization at lower temperatures and contribute with a smaller energy than initially available, as was figured out by Wawszczak et al. [123].

7.1.3. Data conversion

In order to bridge the gap between unstructured finite elements and a regular finite difference grid, the Voronoi tessellation [5] is used to construct a grain structure, as described by Raabe and Becker [90]. A finite element after rolling is a deformed cube which contains eight integration points inside the volume. A representative finite element volume, the size of $30 \times 30 \times 30$ elements, contains 216 000 integration points. Each integration point represents one row, containing grain number, element number, position, the three Euler angles and the stored energy estimation (accumulated plastic strain) of the text file for the data transfer.

To map integration point data onto a regular grid, a Voronoi diagram is created and the finite element integration points are used as Voronoi points. Crystal orientation and stored energy of an integration point are mapped to all discretization points of the regular grid, which belong to the corresponding Voronoi cell. For simplicity, the finite element mesh is not transferred and an interpolation between integration points is not performed, in contrast to the work of Güvenc [34].

The Voronoi point cloud is cut, as indicated by the red box in figure 7.2 a). Cutting is applied to avoid Voronoi artifacts near the simulation domain boundaries because no mesh restricts the Voronoi tessellation. Figure 7.2 b) shows a 2D example of Voronoi artifacts between the black and red squares. Due to this procedure, about 9.5% (20 625) of the 216 000 integration points is cut away to reduce tessellation artifacts.

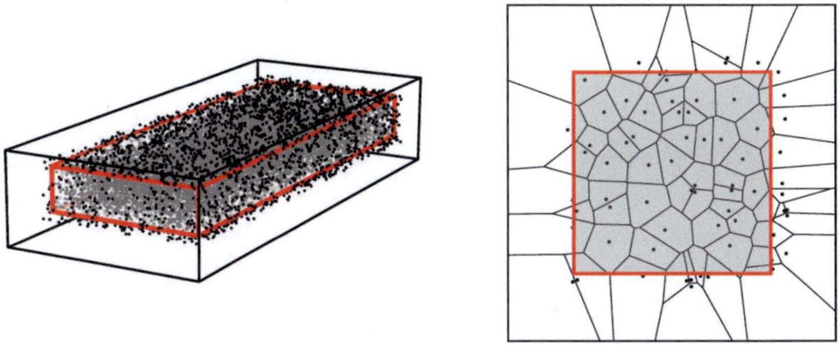

a) b)

Figure 7.2.: Black dots indicate integration points, which are used as Voronoi points to map the finite element data on a finite difference grid. a) Exemplary cutting of the finite element domain. The red domain of the finite element domain (black lines) is mapped on a finite difference grid to avoid tessellation artifacts. b) Examples of Voronoi artifacts (very large Voronoi cells) in 2D are between the black and red squares.

Before constructing the 3D Voronoi diagram, the fineness of the regular grid is scaled to an appropriate value, to ensure sufficient resolution to conduct reliable phase-field simulations, so that a grain has a diameter of about 10 to 16 grid points.

7.1.4. Free energy formulation

The recrystallization process is based on the multiphase and multicomponent phase-field model introduced by Nestler et al. [80] in chapter 3. An additional energy term is introduced to describe the stored energy as the driving force for the recrystallization of grains. To extend the model, the external variable E_{store} for the stored energy of the finite element simulation is used and an additional bulk energy density term $f(\boldsymbol{x}, \boldsymbol{\phi}; E_{\text{store}})$ is formulated in order to model the growth of nuclei.

The driving force for the growth of the nuclei is specified at each discretization point as the stored energy E_{store} from the deformation simulation and $\boldsymbol{\phi}$ by

$$f(\boldsymbol{x}, \boldsymbol{\phi}; E_{\text{store}}) = E_{\text{store}}(\boldsymbol{x}) \sum_{i}^{N} m_i h(\phi_i). \tag{7.3}$$

E_{store} is introduced as an external and position-dependent field to accentuate the static behavior of the field. The factor m_i selects the driving force for each phase i independently and is zero for grains that are not recrystallizing or one for all recrystallizing grains. In addition, it is possible to use $\boldsymbol{m}$ to scale the driving forces. The values of m_i are determined by selecting the nucleation threshold described in section 7.2.2. The recrystallizing grains get an additional driving force contribution related to the stored energy. $h(\phi_i)$ is an interpolation function of the phase fields ϕ_i

$$h(\phi_i) = \phi_i^3 \left(6\phi_i^2 - 15\phi_i + 10\right). \tag{7.4}$$

The evolution equation (3.7) is extended to

$$\begin{aligned} \tau\epsilon \frac{\partial \phi_i}{\partial t} =& \epsilon[\nabla \cdot a_{,\nabla\phi_i}(\boldsymbol{\phi}, \nabla\boldsymbol{\phi}) - a_{,\phi_i}(\boldsymbol{\phi}, \nabla\boldsymbol{\phi})] - \frac{1}{\epsilon} w_{,\phi_i}(\boldsymbol{\phi}) \\ &- f_{,\phi_i}(\boldsymbol{x}, \boldsymbol{\phi}; E_{\text{store}}) - \lambda. \end{aligned} \tag{7.5}$$

To model recrystallization as the dominant process over grain growth, it is necessary to pin the microstructure except for the nuclei. In addition, no description of the stored energy evolution is modeled, such that the grain structure would not match with the stored energy field after a short period of grain growth. The assumption is, the grain structure and the stored energy field do not change significantly during recrystallization, such that the phase-field intrinsic curvature minimization has to be turned off.

Curvature minimization of the interface is removed by introducing the additional term $-|\nabla\phi_i|\nabla\cdot\left(\frac{\nabla\phi_i}{|\nabla\phi_i|}\right)$ from Folch et al. [24] to the evolution equation (7.5). In order to add grain coarsening for recrystallizing grains, the term is extended by the factor $(1-m_i)$, which is appropriate, if m_i is not used for scaling and is only set to either zero or one.

In the recrystallization simulation, a three-dimensional computational domain is considered with up to 216 000 subgrain-like grains. The evolution of N grains is described by N phase-field equations of the form as in equation (7.5). To reduce memory usage, a local order parameter reduction method is applied, as described by Gruber et al. [33], Vedantam and Patnaik [116] or Kim et al. [52]. In order to be able to efficiently compute a large 3D simulation with the message passing interface standard [73] on a high performance computer, the simulation domain is decomposed into three dimensions.

7.2. Parameter Identification

7.2.1. Crystal plasticity parameters

For a statistically representative material behavior, it is necessary to build a unit cell with a sufficient number of grains. A Voronoi algorithm generates a grain structure of the material with 1 000 grains. The second step is to assign the Euler angles from the EBSD measurements to the grains by reducing the data to the number of cells. The reduction is performed according to Eisenlohr and Roters [20].

The discretization of the unit cell, with the morphology and texture information represented by 30^3 elements, is chosen to perform the deformation simulations within a moderate computation time of a few hours. Figure 7.3 shows the undeformed unit cell after mapping the grain structure and figure 7.4 illustrates the deformed unit cell at the final state of the simulation.

A thickness reduction of 63% is achieved by using plane strain compression conditions [104]. Periodic boundary conditions are applied to minimize the constraint effects. This means that the coupling of two equivalent points a and b, that are located on opposite sides of the unit cell with the macroscopic deformation gradient, is described by

$$u_i^b - u_i^a = \overline{F}_{ij}\left(x_{j0}^b - x_{j0}^a\right) - \left(x_{i0}^b - x_{i0}^a\right),$$

where x_{i0}^a and x_{i0}^b indicate the position of a point pair in the non-deformed configuration.

Figure 7.3.: Initial finite element grain morphology.

Figure 7.4.: Deformed morphology.

The uniaxial tension tests of the cold rolled material are the only available data of the material beside the EBSD measurements, such that the simulation parameters are determined with inverse simulations. Temperature and rolling speed of the industrial rolling process are not available. First, a rolling simulation is performed on the cube with a normalized length of 1 and a starting set of parameters. The thickness reduction of 63% is achieved in 0.5 seconds. This correlates to an engineering strain rate of 1.26 per second. Then, a tension test with the deformed material is performed. In the last step, the simulated and experimental tension test data are compared. If the virtual stress-strain diagram does not match with the experimental stress-strain diagram, a different set of parameters is selected manually and the process is repeated until the simulated stress-strain curve fits the experimental stress-strain curve.

The result of the inverse simulation process is described as follows: Elasticity is assumed to be isotropic and for the investigated steel a Young's modulus of 200 000 MPa and a Poisson ratio of $\nu = 0.3$ is imposed. The hardening parameters are considered to be equal for both active slip families {110}<111> and {121}<111>. The parameters for self- and latent-hardening in the Voce hardening matrix are set to 1.2. For the ratio of latent- to self-hardening in the same slip family and also for the ratio of slip planes in other slip families (cf. the definition in [42]), the value is set to 1.2. Table 1 contains the resulting hardening parameters of this inverse simulation. Additionally, the strain rate sensitivity exponent n is set to 28 and the reference strain rate $\dot{a} = 1/s$ is applied for a strain rate insensitive simulation.

At the end of the fitting procedure, the virtual tensile test flow curve matches the experimental profile, as depicted in figure 7.5.

Hardening parameter	**Value [MPa]**
τ_0	95
τ_1	100
θ_0	3,500
θ_1	65

Table 7.1.: Hardening parameters after the inverse simulation procedure

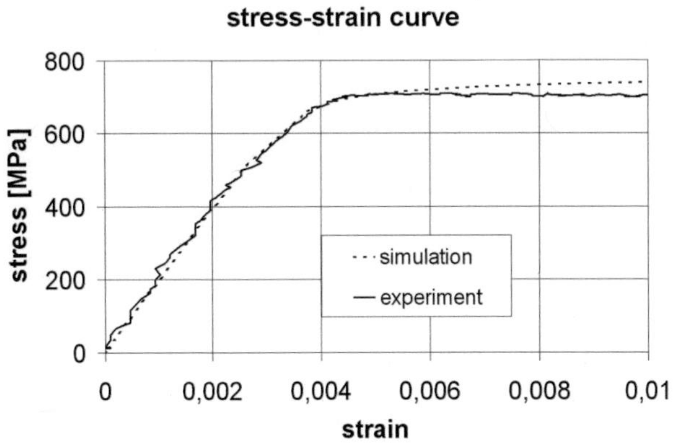

Figure 7.5.: The virtual and experimental stress-strain curves after cold rolling match at the end of the inverse simulation procedure.

7.2.2. Phase-field parameters

It is observed, that subgrains are nuclei or are able to form nuclei for recrystallization, as reviewed by [95] and [16]. Following this line, we define local regions of the converted finite element simulation as nuclei for site-saturated nucleation, so that no new nuclei are introduced during recrystallization. To identify nuclei, we assume that grains with a high amount of stored energy are able to form nuclei.

Two nucleation thresholds based on EBSD data and a user-defined value are introduced, such that two large 3D phase-field simulations are conducted. The threshold based on the EBSD measurements, is defined by the number of grains before and after annealing. Before annealing, 77 088 grains are counted and after annealing, there are 2 565 grains, which is about 3.33%. This value is used to select grains with the highest amount of stored energy and results in 6 505 nuclei (3.33% of 195 375 grains) for the recrystallization simulation. The manually chosen threshold selects the top 0.05% (265 nuclei) grains with the highest amount of stored energy. Figure 7.6 depicts the two nucleation thresholds and the stored energy sorted grains of the finite element simulation.

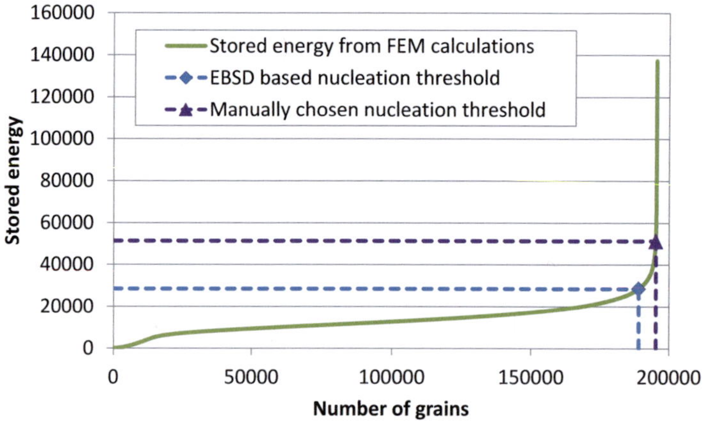

Figure 7.6.: Stored energy sorted grains with the EBSD-based nucleation threshold (top 3.33%) and a manually chosen nucleation threshold (top 0.05%).

Adiabatic conditions are imposed at the domain boundaries for the phase-field method. To simulate recrystallization, we use the following set of dimensionless parameters: Grid spacing is equal in all dimensions and is set to $\Delta x = 1$, the time step width Δt is set to 0.06, the interface thickness parameter is $\epsilon = 5$ and $\gamma_{\alpha\beta\delta}$ is set to the constant value 10.

The total number of time steps is set to 3 000 to capture the recrystallization process for the EBSD-based nucelation threshold simulation, and for the manually chosen nucleation threshold simulation the total number of time steps is set to 11 000, because a higher amount of nuclei will lead to a faster recrystallization. Simulation snapshots are taken after every 100th time step for the EBSD-based threshold simulation and after every 1 000th time step for the manually chosen threshold simulation. In addition to recrystallization, we simulate grain growth for the EBSD-based nucleation threshold simulation until time step 100 000. Because hard disk memory is a limiting factor, we store only the sharp interface field and the stored energy field, where each field requires 1 GB per snapshot. A capacity of about 300 GB will be occupied in total by all snapshots together.

In order to compute the large simulation domain of 1 900×660×214 cells within a reasonable time, 18 469 CPUs are utilized, such that each CPU takes care of about 25×25×25 cells. To decrease computation time further, only cells with a gradient are computed. In other words, a cell update is computed only in the interface area and not in the bulk area.

7.3. Results

Before presenting the orientation distribution functions (ODF) in Euler space after each experimental and simulated process step, the simulation results are compared with the JMAK model and the ODFs after parameterization, recrystallization and grain growth are shown.

To provide a comparision with the JMAK model, we assume site-saturated nucleation with the JMAK exponent of 3 and a shape factor of $\frac{4\pi}{3}$ for spheres, such that the growth rate is used as the fitting parameter to match the simulations. Because the EBSD-based nuclei have a volume fraction of 6% from the start of the simulation, we shift the volume fraction with the dimensionless time of 40 to provide a better match with the

JMAK model. This is not necessary for the manually chosen nucleation threshold volume fractions. Figure 7.7 depicts the fitted JMAK model with two different recrystallization speeds due to a different amount of initial nuclei. The JMAK model shows deviations compared to the simulation results, because the two methods are based on different assumptions.

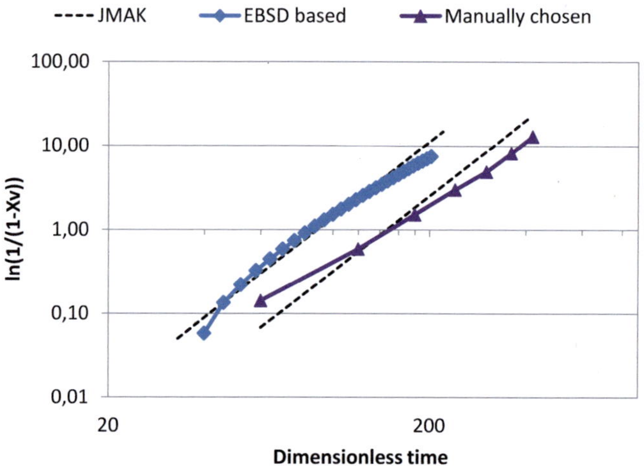

Figure 7.7.: The recrystallized volume fractions X_v are compared with the JMAK fits.

To evaluate the ODFs for body-centered cubic (bcc) steel sheets, it is sufficient to plot the $\phi_2 = 45°$ slice of the Euler space with cubic crystal symmetry and an orthorhombic specimen symmetry. The fibers of interest after cold rolling and annealing are the α fiber ($\phi_1 = 0°$, $0° \leq \Phi \leq 90°$, $\phi_2 = 45°$) and the γ fiber ($60° \leq \phi_1 \leq 90°$, $\Phi = 54.7°$, $\phi_2 = 45°$). Both fibers are visible at $\phi_2 = 45°$ and their development during rolling and annealing is nicely explained by Raabe and Lücke in [91]. The α fiber and the γ fiber develop during cold rolling. During annealing, the α fiber reduces its intensity except for $\Phi = 54.7°$ and the γ fiber increases its intensity further. Figure 7.8 a) depicts the Euler space and figure 7.8 b) depicts the α and γ fibers at $\phi_2 = 45°$.

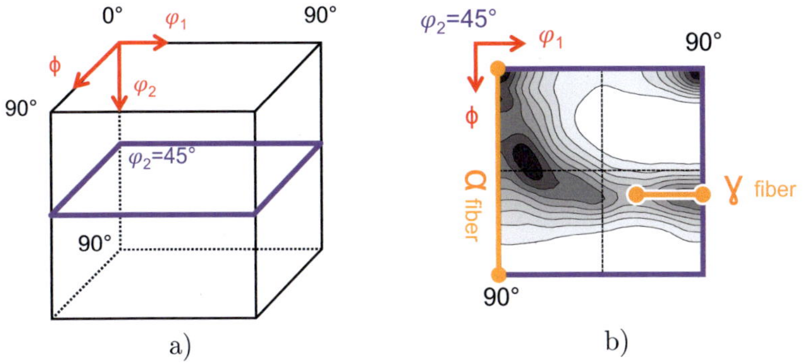

Figure 7.8.: a) The Euler space of a cubic crystal symmetry and an orthorhombic specimen symmetry with a slice at $\phi_2 = 45°$. b) The location of the α and γ fibers at the $\phi_2 = 45°$ slice with the experimentally measured ODF after cold rolling in the background.

It is possible to estimate the recrystallization ODF by assuming an equal volume fraction of the nuclei after recrystallization. In figure 7.9, these estimations are compared with the ODFs after numerical recrystallization, together with the orientation distribution of the last snapshot for the EBSD-based nucleation threshold simulation. All ODFs in figure 7.9 have a strong γ fiber in common. The maximum value difference between the estimated and the recrystallized ODFs is significantly higher for the manually chosen nucleation threshold simulation, which is accompanied by a bigger change of the toal distribution. In contrast to the manually chosen nucleation threshold, the recrystallization process does not change the ODF significantly for the EBSD-based nucleation threshold simulation.

In figure 7.10, the $\phi_2 = 45°$ slices of the ODFs are ordered along the process chain according to the schematic drawing in figure 7.1. The experimental ODF consists of 8 416 orientations and the ODF of the finite element starting condition covers 1 000 orientations after hot rolling. Because only a part of the measured orientations is used to generate the starting conditions for the finite element computantion, the ODFs are not equal. But both ODFs are nearly random, which is indicated by the maximum value around 1.

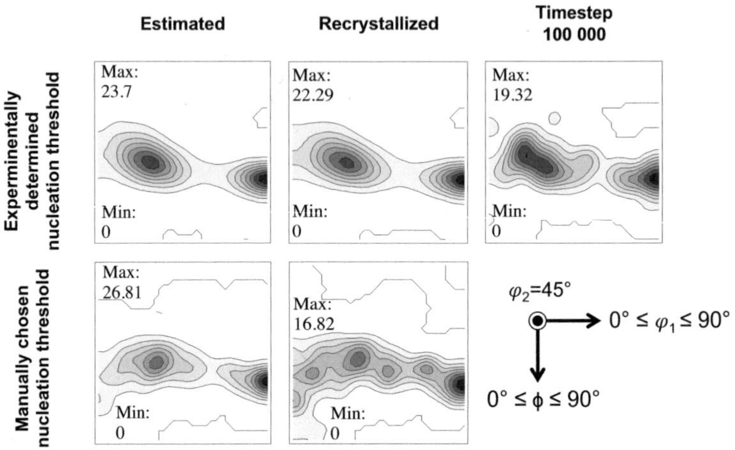

Figure 7.9.: The rows contain the ODFs for the EBSD-based and manually chosen nucleation thresholds. The first column contains the estimated ODFs based on the nucleation thresholds. The second column contains the ODFs of the numerically recrystallized grain structure. The last column shows the ODF of the last simulated time step.

After cold rolling, the development of the α and γ fibers is reproduced by the crystal plasticity model, as well as by the experiment. The maximum value produced by the simulation is higher than in the experiment and the peaks are located reasonably close to the experimental measurements. Another difference between experiment and simulation is the pronounced α fiber in the experiment and the strong γ fiber of the simulation, as well as the strong texture components {001}<110> ($\phi_1 = 45°$, $\Phi = 0°$, $\phi_2 = 45°$) and {001}<010> ($\phi_1 = 90°$, $\Phi = 0°$, $\phi_2 = 45°$) in the experiment. The deviations can be related to the small sample size of 1 000 low resolution grains.

The data conversion step between the finite elements and the finite difference grid does not influence the ODF significantly, although 9.5% of the integration points are deleted. The ODF of the manually chosen nucleation threshold is used in figure 7.10 to represent the recrystallized state. After annealing, the γ fiber is strengthened, the α fiber intensities are reduced and the maximum value is increased in the experiment, as well as in the simulation.

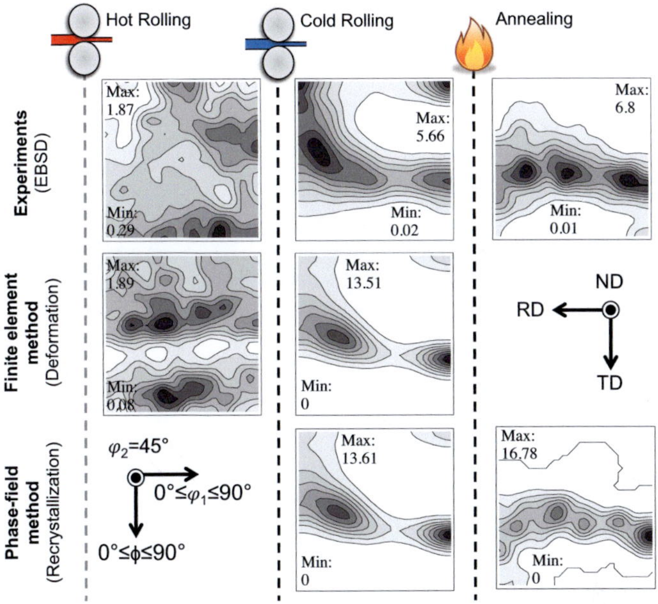

Figure 7.10.: The simulations are able to reproduce the major developments of the α and the γ fibers compared to the experiments. The differences after hot rolling are neglecable due to nearly random orientation distributions. The slices through the Euler space at $\phi_2 = 45°$ of the ODFs after each process step for the experiments, the finite element method and the phase-field method layer are sufficient for comparison of body-centered cubic metals after cold rolling and annealing.

To complete the results section, the grain structures in figure 7.11 after data conversion, recrystallization and grain growth is presented, where inverse pole figure coloring is used to show the crystal lattice orientations.

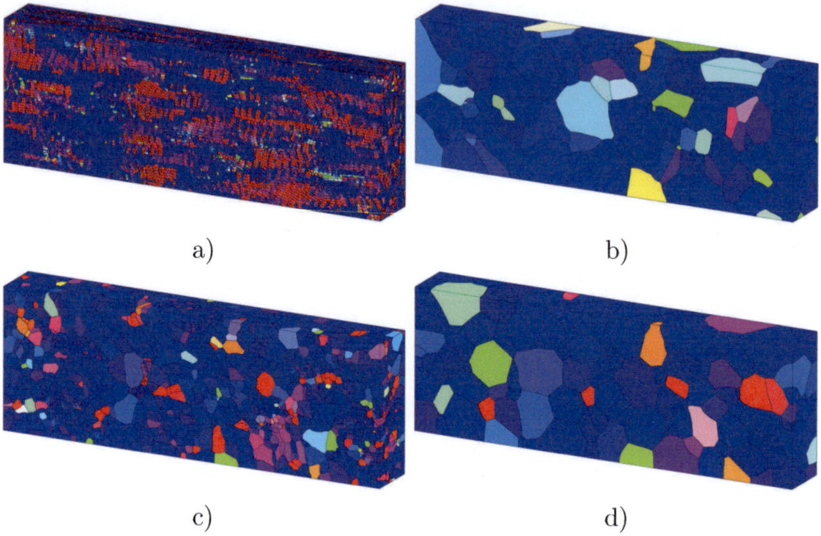

Figure 7.11.: Inverse pole figure colored grain structures a) after data conversion, b) after recrystallization with the manually chosen nucleation threshold, c) after recrystallization with the EBSD-based nucleation threshold and d) after some grain growth of the EBSD-based nucleation threshold simulations.

7.4. Discussion

In contrast to the work of Raabe and Becker [90] and Takaki and Tomita [112], an experimentally accompanied, coupled approach to simulate static recrystallization in 3D with many grains or subgrain-like grains is presented, which allows a more reliable analysis of the orientation distributions. Unfortunately, the larger scale prevents a detailed description of the subgrain structure, as presented by Suwa et al. [110] and Takaki and Tomita [112]. The subgrain structure is critical to reproduce the nucleation process properly, as reviewed by Rios et al. [95].

The crystal plasticity finite element model is able to reproduce the major crystal orientation changes, together with a reasonable estimation of stored energy. Incorporating a higher resolution of the grains during the deformation process or using a more precise description of the dislocation distribution by crystal dislocation theory [57], continuum dislocation dynamics theory [101] or even discrete dislocation dynamics, would increase the validity of the starting condition for recrystallization. On the other hand, the computational costs would increase, the higher the level of detail.

The overall approach seems promising to investigate recrystallization phenomena, but further refinements are necessary to reproduce the orientation distributions during deformation more accurately and to select the nuclei more sophisticatedly such as temperature-dependent material quantities. One of the next questions is, whether the same approach is able to produce similar results for face-centered cubic metals, where other fibers and texture components are of interest.

7.5. Conclusion

We present a coupled approach of a crystal plastiticy finite element model with a phase-field model to simulate the grain structure evolution during combined cold rolling (deformation) and annealing (static recrystallization) processes for DC04 steel in a multiscale 3D treatment. The simulations are supported by experimental measurements to parameterize the simulations and to validate the results. Attention is placed on the orientation distributions after each process step, where both simulation methods are able to capture the major developments of the α and γ fibers during cold rolling and annealing, such that the simulation results are in good agreement with the experiments and the literature.

The used crystal plasticity model is able to produce reasonable estimations of stored energy and crystal lattice rotations for the simulation of recrystallization in 3D. Without resolving a complete subgrain structure, the simplified nucleation process of selecting the nuclei by an energy threshold, delivers appropriate results by means of orientation distributions. Depending on the nucleation threshold, it can be necessary to simulate recrystallization to capture texture changes. Furthermore, a seamless transition from recrystallization to grain growth is achieved. With a high performance computation facility, the large recrystallization simulations are computed within reasonable time.

8. Outlook

The study of regular grains in 2D with different grain boundary energies showed the isotropic behavior in average, but the growth rate distributions show a variety of possible growth rates without directly abnormal growing grains. A study of the topological arrangements incorporating the next neighbors will be a direct next step to relate abnormal grain growth with grain boundary energy anisotropy and to compare the model of Humphreys in [44].

To incorporate curvature and to simulate a longer timespan with the vertex model is a very interesting and important refinement. Further, a vertex model with faceted anisotropy could reveal the secret behind the abnormally fast growing grains.

3D simulations are a large computational challenge for the brute force approach, even with a fast vertex model. The parameter space is growing exponentially, such that only the largest computation facilities would be able to compute such a challenging task.

The communication bound, parallel computation of large 2D and 3D domains makes a 2D decomposition and a 3D decomposition, respectively, necessary to be able to compute large-scale simulations. The clear performance model based on a timing model is able to support the necessity of higher order decomposition to achieve a sufficient efficiency. In order to achieve an even better performance, it is necessary to perform a detailed analysis of the memory usage. One step further would be to implement 3D or 3.5D blocking to minimize cache misses [82] and to improve the sequential runtime. 3D load balancing, adaptive mesh refinement and vectorization would reduce the computation time to a minimum.

The numerical recrystallization of body-centered cubic iron with the coupled approach was able to capture major texture development processes.

A simulation series with different thickness reductions together with a parameter study for nucleation would be able to identify strengths and weaknesses of the presented approach. It would be of great interest, if the current coupled approach is able to produce good results for face-centered cubic metals in the same process chain. For fcc metals, the fibers of interest and the development is different.

One common tendency can be derived from this work. New opportunities to gain insights and to identify or validate relations are enabled by increasing the computational scale. The results of large-scale investigations often involve large amounts of data, which have to be handled with ease to be able to analyze or visualize them. Furthermore, the way of increasing the scale to the limit or to the next order of magnitude is a scientific task of its own, because scientific computing is diverse.

A. Appendix

Grain growth behavior

The following diagrams are related to chapter 3 Grain growth behavior. The diagrams show the growth rates of the n-sided grain simulation series.

The growth rates (blue diamonds) and the average growth rates (red square) of the 500 n-sided grain series with the Read–Shockley and faceted anisotropy models for $4 \leq n \leq 8$.

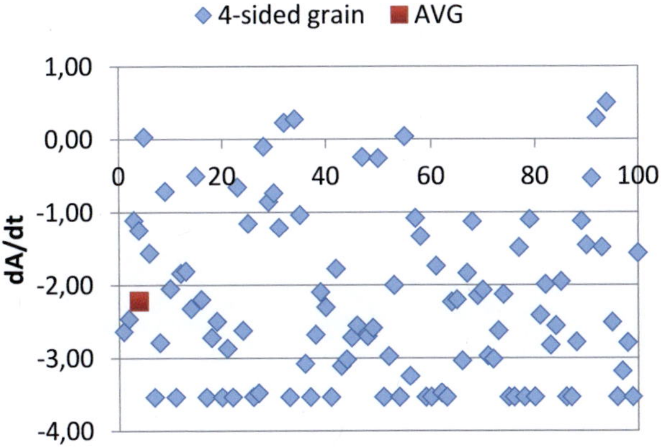

Figure A.1.: Growth rates (blue diamonds) with the average value (red square) of a 4-sided grain for 100 random lattice configurations with the Read–Shockley model. Shrinkage with smaller than 3.5 is not captured, because the central grain had a too small area.

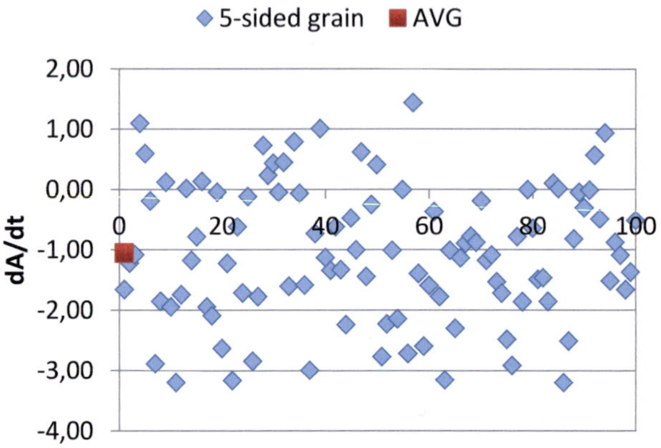

Figure A.2.: Growth rates (blue diamonds) with the average value (red square) of a 5-sided grain for 100 random lattice configurations with the Read–Shockley model.

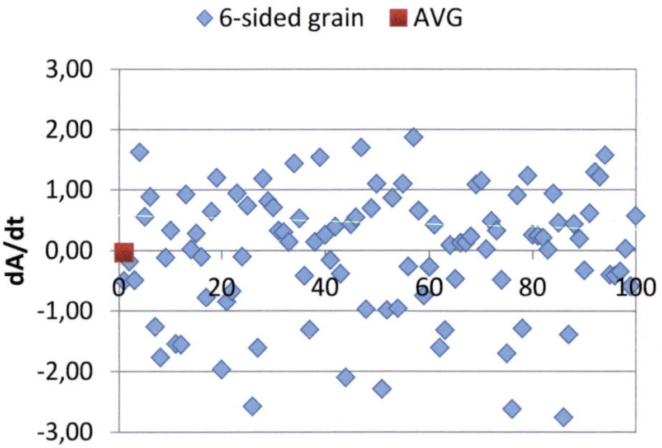

Figure A.3.: Growth rates (blue diamonds) with the average value (red square) of a 6-sided grain for 100 random lattice configurations with the Read–Shockley model.

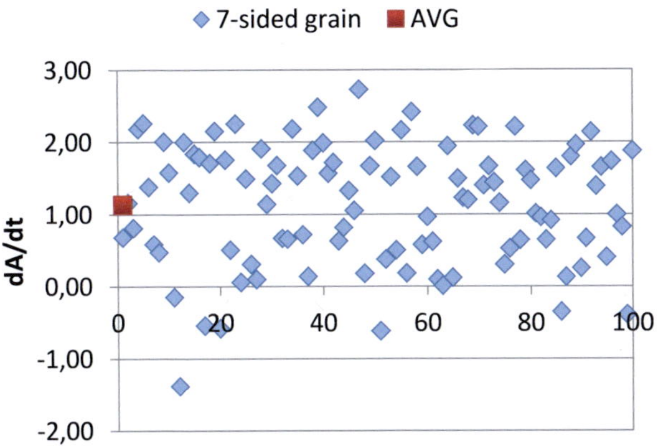

Figure A.4.: Growth rates (blue diamonds) with the average value (red square) of a 7-sided grain for 100 random lattice configurations with the Read–Shockley model.

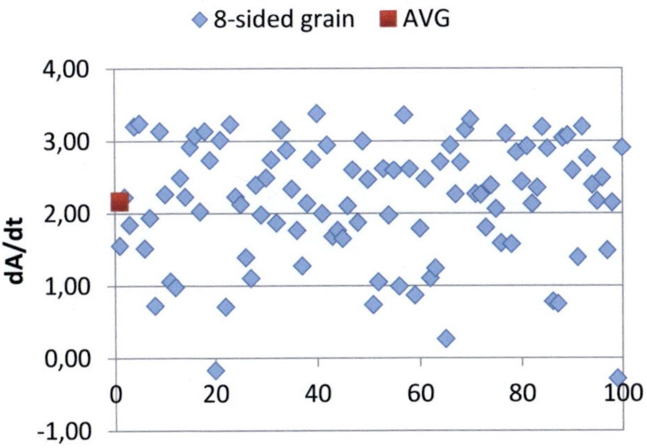

Figure A.5.: Growth rates (blue diamonds) with the average value (red square) of a 8-sided grain for 100 random lattice configurations with the Read–Shockley model.

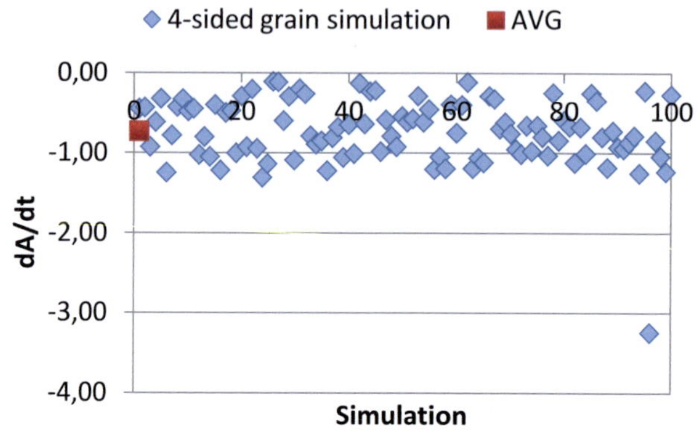

Figure A.6.: Growth rates (blue diamonds) with the average value (red square) of a 4-sided grain for 100 random lattice configurations with the faceted anisotropy model. One outlier is present with a negative growth rate

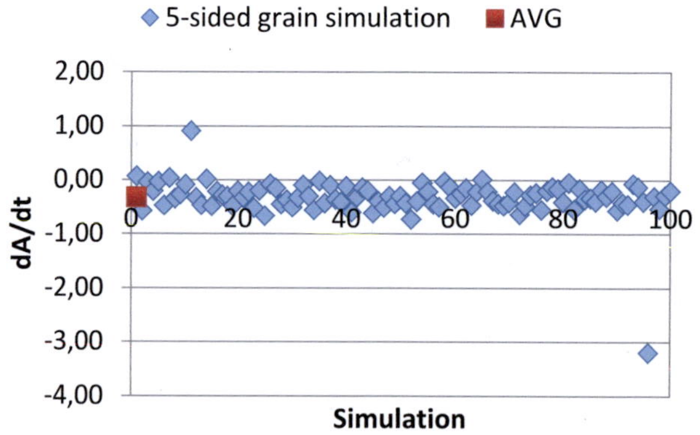

Figure A.7.: Growth rates (blue diamonds) with the average value (red square) of a 5-sided grain for 100 random lattice configurations with the faceted anisotropy model. One outlier is present with a negative growth rate and one outlier has a positive growth rate, whereas most configurations lead to negative growth rates.

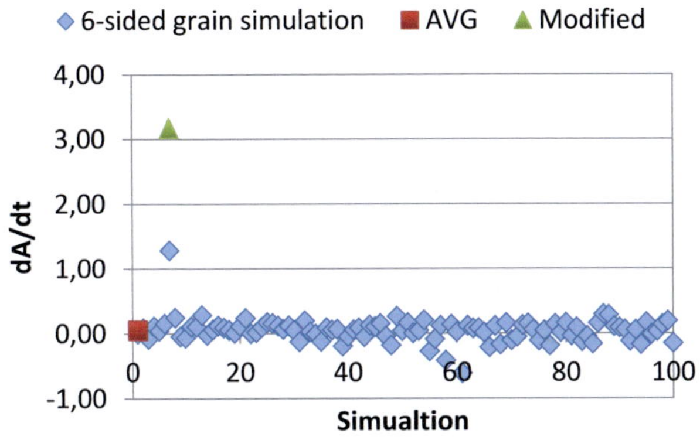

Figure A.8.: Growth rates (blue diamonds) with the average value (red square) of a 6-sided grain for 100 random lattice configurations with the faceted anisotropy model. The configuration, which leads to a positive growth rate is modified, such that an even faster growth (green triangle) is obtained.

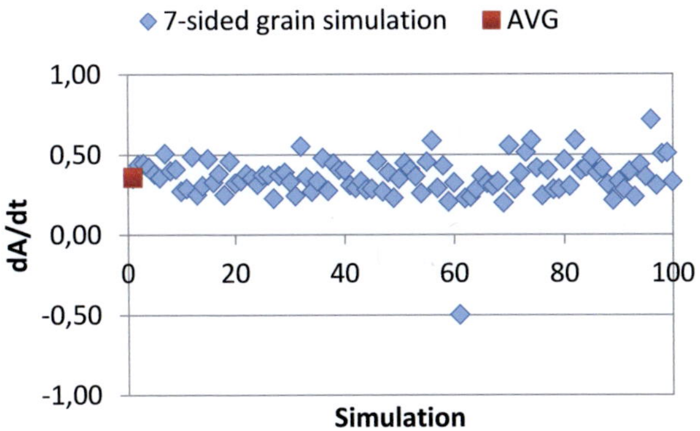

Figure A.9.: Growth rates (blue diamonds) with the average value (red square) of a 7-sided grain for 100 random lattice configurations with the faceted anisotropy model. One configuration leads to a fast shrinkage.

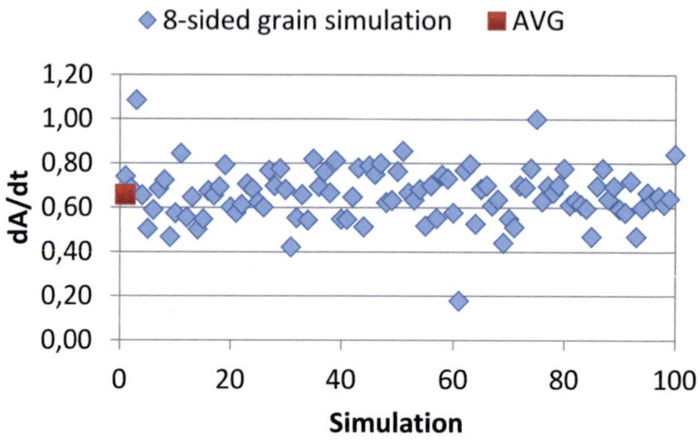

Figure A.10.: Growth rates (blue diamonds) with the average value (red square) of a 8-sided grain for 100 random lattice configurations with the faceted anisotropy model. A clear outlier is not present but some growth rates are significantly above or beneath

Performance measurements

All runtime measurements of the simulation settings A, B and C from chapter 5 and 6 are listed here.

CPUs	Seconds	Speedup	Efficiency
1	254 997	-	100.00%
7	41 116	6.20	88.60%
15	20 197	12.63	84.17%
31	10 951	23.29	75.11%
63	5 324	47.90	76.03%
127	3 116	81.83	64.44%
255	1 979	128.85	50.53%

Table A.1.: Runtime, speedup and efficiency of the 1 000×1 000 cells 2D dendrite with 1D domain decomposition (configuration A.1).

CPUs	Seconds	Speedup	Efficiency
1	2082934	-	100,00%
8	316289	6,59	82,32%
16	169391	12,30	76,85%
32	93247	22,34	69,81%
64	57337	36,33	56,76%
128	37043	56,23	43,93%
200	33040	63,04	31,52%

Table A.2.: Runtime, speedup and efficiency of the 200×200×200 cells 3D dendrite with 1D domain decomposition (configuration A.2).

CPUs	Seconds	Speedup	Efficiency
1	2082 934	-	100.00%
8	219 562	9.49	118.58%
16	134 177	15.52	97.02%
32	60 072	34.67	108.36%
64	34 626	60.16	93.99%
128	19 669	105.90	82.73%
256	9 580	217.43	84.93%

Table A.3.: Runtime, speedup and efficiency of the 200×200×200 cells 3D dendrite with 3D domain decomposition (configuration A.2).

CPUs	Seconds	Speedup	Efficiency
1	878 731	-	100.00%
7	162 430	5.41	77.28%
15	85 039	10.33	68.89%
31	47 032	18.68	60.27%
63	29 679	29.61	47.00%
127	22 802	38.54	30.34%

Table A.4.: Runtime, speedup and efficiency of the 500×500×500 cells 3D dendrite with 1D domain decomposition (configuration A.3).

Growth rate distributions

The growth rate distributions of chapter 4 for regular n-sided grains with $3 \leq n \leq 8$ are computed with the vertex model. The n-sided grains are symmetrically but the resulting growth rate distributions are not symmetric and move from negative growth rates to positive growth rates with increasing neighbor count.

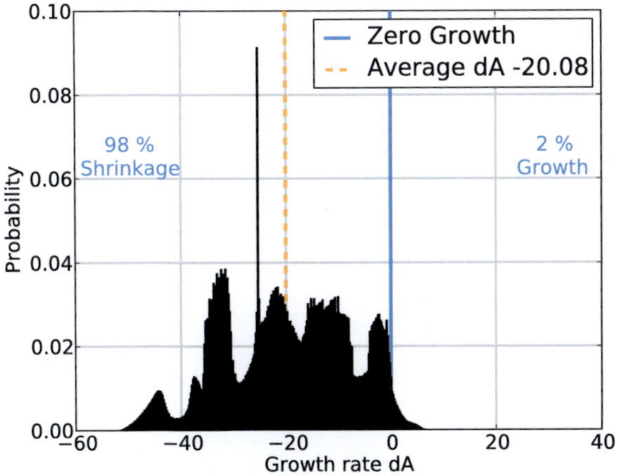

Figure A.11.: The growth rate distribution of a 3-sided grain is characterized by two small and four large cusps in the negative growth regime. The isotropic peak is pronounced compared to the growth rate distributions for more than three neighbors.

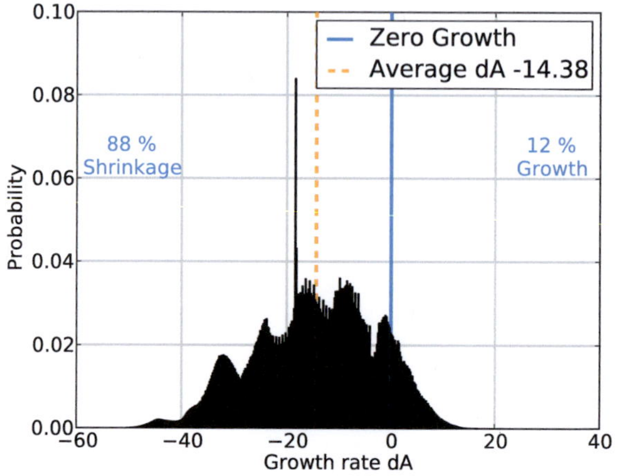

Figure A.12.: The growth rate distribution of a 4-sided grain is characterized by two small and five large cusps.

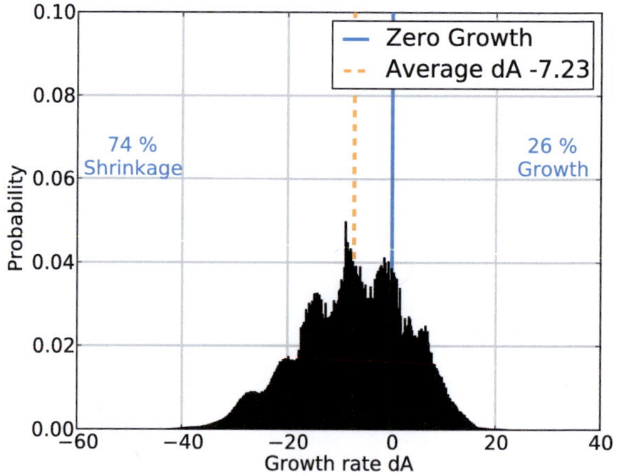

Figure A.13.: The growth rate distribution of a 5-sided grain is characterized by six superimposed cusps. Almost one third of the growth rates are in the positive growth regime.

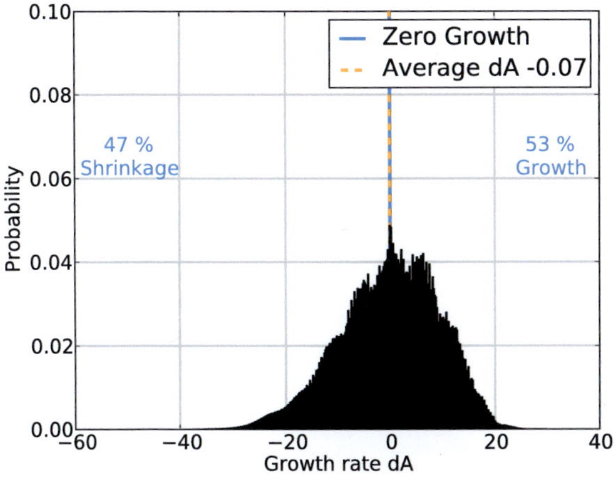

Figure A.14.: The growth rate distribution of a 6-sided grain is characterized by an average growth rate of zero and no distinct cusp.

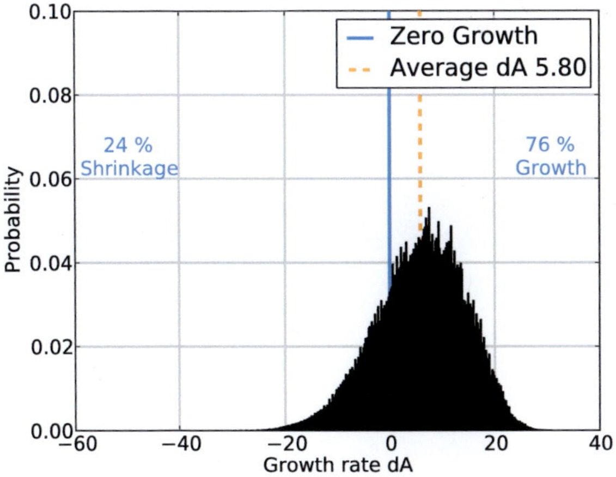

Figure A.15.: The growth rate distribution of a 7-sided grain aligns to a normal distribution.

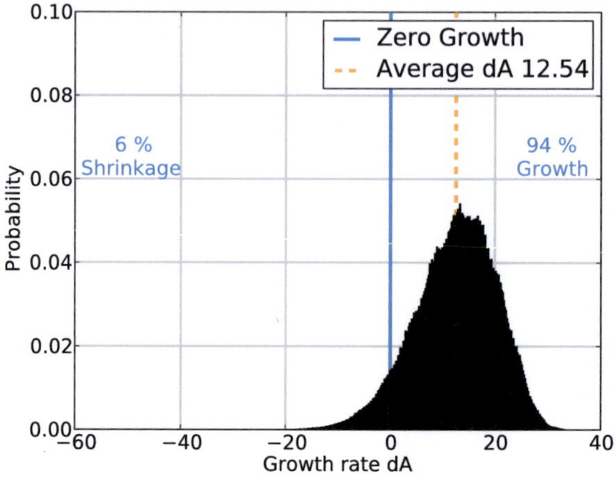

Figure A.16.: The growth rate distribution of a 8-sided grain consists mainly of growing configurations.

Tools

List of tools used to perform this work.

Simulation	
Simulation	PACE3D
Analysis	PACE3D, Excel, bash, sed, awk, Mathematica, Python
ODF analysis	Matlab with MTEX 4.0
Programming	
Programming	Eclipse, QT
Debugging	valgrind, gdb
Parallelization	MPI, OpenMP
Parallel Debugging	Allinea Distributed Debugging Tool
Version Control	git
Art and Text	
Illustrating	PowerPoint, SketchUp, KolourPaint, Gimp, Python, Mathematica, PACE3D, TikZ, Gnuplot
Text editing	Kate, Sublime Text, vi, nano, Kile
Typesetting	LaTeX
Organization	
Reference Manager	Mendeley, JabRef
Office	Microsoft Office, Libre Office, Open Office
Collaboration	Dropbox, F*EX file exchange, USB Drive, Google Hangouts, Adium, Thunderbird, Pidgin
Miscellaneous	
Browser	Mozilla Firefox, Google Chrome, Konqueror
Language	C, C++, bash, Python, Visual Basic, OpenCL

Bibliography

[1] H. Abdolvand, M.R. Daymond, and C. Mareau. Incorporation of twinning into a crystal plasticity finite element model: Evolution of lattice strains and texture in zircaloy-2. International Journal of Plasticity, 2011.

[2] R.J. Asaro. Crystal Plasticity. Journal of Applied Mechanics, 50:921–934, 1983.

[3] R.J. Asaro. Micromechanics of crystals and polycrystals, volume Volume 23 of Advances in Applied Mechanics, pages 1–115. Elsevier, 1983.

[4] World Steel Associaton. Steel statistical yearbook 2011. Technical report, World Steel Associaton, 2011.

[5] F. Aurenhammer. Voronoi diagrams - a survey of a fundamental geometric data structure. ACM Comput. Surv., 23:345–405, 1991.

[6] M. Avrami. Kinetics of Phase Change. I General Theory. The Journal of Chemical Physics, 7:1103, 1939.

[7] F. Bachmann, R. Hielscher, and H. Schaeben. Texture analysis with mtex–free and open source software toolbox. Solid State Phenomena, 160:63–68, 2010.

[8] M. Balzer and O. Deussen. Voronoi treemaps. In IEEE Symposium on Information Visualization, INFOVIS 2005., pages 49–56. IEEE, 2005.

[9] K. Barmak, E. Eggeling, D. Kinderlehrer, R. Sharp, S. Ta'Asan, A.D. Rollett, and K.R. Coffey. Grain growth and the puzzle of its stagnation in thin films: The curious tale of a tail and an ear. Progress in Materials Science, 58:987–1055, 2013.

[10] M. Bäurer, M. Syha, and D. Weygand. Combined experimental and numerical study on the effective grain growth dynamics in highly anisotropic systems: Application to barium titanate. Acta Materialia, 61:5664–5673, 2013.

[11] C. Bergmann. Über die Verhältnisse der Wärmeökonomie der Thiere zu ihrer Grösse. Vandenhoeck und Ruprecht, Göttingen, göttinger edition, 1848.

[12] L.Q. Chen. Phase-Field Models For Microstructure Evolution. Annual Review of Materials Research, 32:113–140, 2002.

[13] Y.P. Chen, W.B. Lee, and S. To. Influence of initial texture on formability of aluminum sheet metal by crystal plasticity fe simulation. Journal of materials processing technology, 192:397–403, 2007.

[14] M. Crumbach, M. Goerdeler, G. Gottstein, L. Neumann, H. Aretz, and R. Kopp. Through-process texture modelling of aluminium alloys. Modelling and Simulation in Materials Science and Engineering, 12:S1–S18, 2004.

[15] S.J. Dillon, M. Tang, W.C. Carter, and M.P. Harmer. Complexion: A new concept for kinetic engineering in materials science. Acta Materialia, 55:6208–6218, 2007.

[16] R. Doherty, D. Hughes, F. Humphreys, J. Jonas, D. Jensen, M. Kassner, W. King, T. Mcnelley, H. Mcqueen, and A. Rollett. Current issues in recrystallization: a review. Materials Science and Engineering A, 238:219–274, 1997.

[17] J. Dongarra, P. Beckman, P. Aerts, F. Cappello, T. Lippert, S. Matsuoka, P. Messina, T. Moore, R. Stevens, A. Trefethen, and M. Valero. The International Exascale Software Project: a Call To Cooperative Action By the Global High-Performance Community. International Journal of High Performance Computing Applications, 23:309–322, 2009.

[18] J. Dongarra, I. Foster, G. Fox, W. Gropp, K. Kennedy, L. Torczen, and A. White. Sourcebook of parallel computing. Morgan Kaufmann, 2009.

[19] J.J. Eggleston, G.B. McFadden, and P.W. Voorhees. A phase-field model for highly anisotropic interfacial energy. Physica D: Nonlinear Phenomena, 150:91–103, 2001.

[20] P. Eisenlohr and F. Roters. Selecting a set of discrete orientations for accurate texture reconstruction. Computational Materials Science, 42:670–678, 2008.

[21] P. Eisenlohr, D.D. Tjahjanto, T. Hochrainer, F. Roters, and D. Raabe. Comparison of texture evolution in fcc metals predicted by various grain cluster homogenization schemes. International Journal Of Materials Research, 100:500–509, 2009.

[22] M. Elsey, S. Esedoglu, and P. Smereka. Simulations of anisotropic grain growth: Efficient algorithms and misorientation distributions. Acta Materialia, 61:2033–2043, 2013.

[23] V. Faber, O.M. Lubeck, and A.B. White. Superlinear speedup of an efficient sequential algorithm is not possible. Parallel Computing, 3:259–260, 1986.

[24] R. Folch, J. Casademunt, A. Hernández-Machado, and L. Ramírez-Piscina. Phase-field model for Hele-Shaw flows with arbitrary viscosity contrast. I. Theoretical approach. Physical review. E, Statistical physics, plasmas, fluids, and related interdisciplinary topics, 60:1724–33, 1999.

[25] V.E. Fradkov and D. Udler. Two-dimensional normal grain growth: topological aspects. Advances in Physics, 43:739–789, 1994.

[26] K. Fuchizaki, T. Kusaba, and K. Kawasaki. Computer modelling of three-dimensional cellular pattern growth. Philosophical Magazine Part B, 71:333–357, 1995.

[27] M. Geimer, F. Wolf, B.J.N. Wylie, E. Ábrahám, D. Becker, and B. Mohr. The Scalasca performance toolset architecture. Concurr. Comput. : Pract. Exper., 22:702–719, 2010.

[28] W.L. George and J.A. Warren. A parallel 3D dendritic growth simulator using the phase-field method. Journal of Computational physics, 177:264–283, 2002.

[29] OR. Gerard. A brief history of computing, 2012.

[30] R. González-Cinca, L. Ramírez-Piscina, J. Casademunt, A. Hernández-Machado, L. Kramer, T. Tóth Katona, T. Börzsönyi, and Á. Buka. Phase-field simulations and experiments of faceted growth in liquid crystals. Physica D: Nonlinear Phenomena, 99:359–368, 1996.

[31] G. Gottstein, Y. Ma, and L. Shvindlerman. Triple junction motion and grain microstructure evolution. Acta Materialia, 53:1535–1544, 2005.

[32] G.S. Grest, D.J. Srolovitz, and M.P. Anderson. Computer simulation of grain growth - IV. Anisotropic grain boundary energies. Acta Metallurgica, 33:509–520, 1985.

[33] J. Gruber, N. Ma, Y. Wang, A.D. Rollett, and G.S. Rohrer. Sparse data structure and algorithm for the phase field method. Modelling and Simulation in Materials Science and Engineering, 14:1189–1195, 2006.

[34] O. Güvenc, T. Henke, and G. Laschet. Modeling of static recrystallization kinetics by coupling crystal plasticity FEM and multiphase field calculations. Computer Methods in Materials Science, 13:368–374, 2013.

[35] P. Haupt. Continuum Mechanics and Theory of Materials. Springer, 2002.

[36] D. Helm. Thermomechanical representation of the stored energy during plastic deformation. International Journal of Materials Research, 101:972–980, 2010.

[37] D. Helm, A. Butz, D. Raabe, and P. Gumbsch. Microstructure-based description of the deformation of metals: Theory and application. JOM Journal of the Minerals, Metals and Materials Society, 63:26–33, 2011.

[38] D.P. Helmbold. Modelling speedup (n) greater than n. Parallel and Distributed, 1:250–256, 1990.

[39] C. Herring. The Physics of Powder Metallurgy, pages 143–156, 1951.

[40] C. Herring. Some theorems on the free energies of crystal surfaces. Physical Review, 82:87–93, 1951.

[41] E.A. Holm and C.C. Battaile. The computer simulation of microstructural evolution. Jom, 53:20–23, 2001.

[42] Y. Huang. A user-material subroutine incorporating single crystal plasticity in the ABAQUS finite element program. Mech Report, 178, 1991.

[43] R.E. Hummel. Understanding Materials Science: History, Properties, Applications, Second Edition. Springer, 2004.

[44] F.J. Humphreys. A unified theory of recovery, recrystallization and grain growth, based on the stability and growth of cellular microstructures i. the basic model. Acta Materialia, 45:4231–4240, 1997.

[45] F.J. Humphreys. A unified theory of recovery, recrystallization and grain growth, based on the stability and growth of cellular microstructures ii. the effect of second-phase particles. Acta Materialia, 45:5031–5039, 1997.

[46] J.W. Hutchinson. Elastic-plastic behaviour of polycrystalline metals and composites. Proceedings of the Royal Society A Mathematical Physical and Engineering Sciences, 319:247–272, 1970.

[47] A.E. Johnson and P.W. Voorhees. A phase-field model for grain growth with trijunction drag. Acta Materialia, 67:134–144, 2014.

[48] K. Kawasaki, T. Nagai, and K. Nakashima. Vertex models for two-dimensional grain growth. Philosophical Magazine Part B, 60:399–421, 1989.

[49] A. Kazaryan, B.R. Patton, S.A. Dregia, and Y. Wang. On the theory of grain growth in systems with anisotropic boundary mobility. Acta Materialia, 50:499–510, 2002.

[50] A. Kazaryan, Y. Wang, S. Dregia, and B. Patton. Grain growth in systems with anisotropic boundary mobility: Analytical model and computer simulation. Physical Review B, 63:184102, 2001.

[51] A. Kazaryan, Y. Wang, S.A. Dregia, and B.R. Patton. Grain growth in anisotropic systems: comparison of effects of energy and mobility. Acta Materialia, 50:2491–2502, 2002.

[52] S.G. Kim, D.I. Kim, W.T. Kim, and Y.B. Park. Computer simulations of two-dimensional and three-dimensional ideal grain growth. Physical Review E - Statistical, Nonlinear and Soft Matter Physics, 74:061605, 2006.

[53] C.E. Krill and L.Q. Chen. Computer simulation of 3-D grain growth using a phase-field model. Acta Materialia, 50:3059–3075, 2002.

[54] E. Kröner. Allgemeine Kontinuumstheorie der Versetzungen und Eigenspannungen. Archive for Rational Mechanics and Analysis, 4:273–334, 1959.

[55] Y.J. Lan and C. Pinna. Modelling textures formed during the plane strain compression and subsequent static recrystallisation of Body-Centred Cubic (BCC) metals. Materials Science Forum, 709:3040–3045, 2012.

[56] J.S. Langer. Models of pattern formation in first order phase transitions in directions in condensed matter. World Scientific Singapore, page 165, 1986.

[57] K.C. Le and C. Günther. Nonlinear continuum dislocation theory revisited. International Journal of Plasticity, 53:164–178, 2014.

[58] R.A. Lebensohn and C.N. Tome. A self-consistent anisotropic approach for the simulation of plastic deformation and texture development of polycrystals: Application to zirconium alloys. Acta Metallurgica et Materialia, 41:2611–2624, 1993.

[59] B.K. Lee, S.Y. Chung, and S.J.L. Kang. Grain boundary faceting and abnormal grain growth in BaTiO 3. Acta materialia, 48:1575–1580, 2000.

[60] E.H. Lee and D.T. Liu. Finite-strain elastic–plastic theory with application to plane-wave analysis. Journal of Applied Physics, 38:19–27, 1967.

[61] C.E. Leiserson. Fat-trees: Universal networks for hardware-efficient supercomputing. IEEE Transactions on Computers, C-34:892–901, 1985.

[62] H. Li, C. Wu, and H. Yang. Crystal plasticity modeling of the dynamic recrystallization of two-phase titanium alloys during isothermal processing. International Journal of Plasticity, 51:271–291, 2013.

[63] L. Li and A. Malony. Model-Based Performance Diagnosis of Master-Worker Parallel Computations. In Wolfgang Nagel, Wolfgang Walter, and Wolfgang Lehner, editors, Euro-Par 2006 Parallel Processing, volume 4128 of Lecture Notes in Computer Science, pages 35–46. Springer Berlin / Heidelberg, 2006.

[64] H.K. Lin, H.Y. Chen, and C.W. Lan. Phase field modeling of facet formation during directional solidification of silicon film. Journal of Crystal Growth, 385:134–139, 2014.

[65] Z.L. Liu, X.M. Liu, Z. Zhuang, and X.C. You. A multi-scale computational model of crystal plasticity at submicron-to-nanometer scales. International Journal of Plasticity, 25:1436–1455, 2009.

[66] M.A. Lovette, A.R. Browning, D.W. Griffin, J.P. Sizemore, R.C. Snyder, and M.F. Doherty. Crystal Shape Engineering. Industrial & Engineering Chemistry Research, 47:9812–9833, 2008.

[67] A. Ma, F. Roters, and D. Raabe. A dislocation density based constitutive model for crystal plasticity FEM including geometrically necessary dislocations. Acta Materialia, 54:2169–2179, 2006.

[68] N. Ma, Q. Chen, and Y. Wang. Implementation of high interfacial energy anisotropy in phase field simulations. Scripta Materialia, 54:1919–1924, 2006.

[69] R.D. MacPherson and D.J. Srolovitz. The von Neumann relation generalized to coarsening of three-dimensional microstructures. Nature, 446:1053–5, 2007.

[70] J. Mandel. Plasticite classique et viscoplasticite. Springer, 1973.

[71] G. McFadden, A. Wheeler, R. Braun, S. Coriell, and R. Sekerka. Phase-field models for anisotropic interfaces. Physical review. E, Statistical physics, plasmas, fluids, and related interdisciplinary topics, 48:2016–2024, 1993.

[72] Message Passing Interface Forum. MPI: A Message-Passing Interface Standard Version 2.2. High-Performance Computing Center Stuttgart, Stuttgart, 2009.

[73] Message Passing Interface Forum. MPI : A Message-Passing Interface Standard Version 3.0. High-Performance Computing Center Stuttgart, Stuttgart, 2012.

[74] H.S. Meuer, J. Dongarra, E. Strohmaier, and H. Simon. Top500. http://www.top500.org/, 2014. [Online; accessed 12-March-20014].

[75] M. Miodownik. A review of microstructural computer models used to simulate grain growth and recrystallisation in aluminium alloys. Journal of Light Metals, 2:125–135, 2002.

[76] G.E. Moore. Cramming more components onto integrated circuits. Proceedings of the IEEE, 38, 1965.

[77] M.S. Muller, A. Knüpfer, M. Jurenz, M. Lieber, H. Brunst, H. Mix, and W.E. Nagel. Developing Scalable Applications with Vampir, VampirServer and VampirTrace. Parallel Computing: Architectures, Algorithms and Applications, 38:637–644, 2007.

[78] W.W. Mullins. Two-Dimensional Motion of Idealized Grain Boundaries. Journal of Applied Physics, 27:900, 1956.

[79] M. Muramatsu, Y. Tadano, and K. Shizawa. A Phase-Field Simulation of Nucleation from Subgrain and Grain Growth in Static Recrystallization. Materials Science Forum, 584-586:1045–1050, 2008.

[80] B. Nestler, H. Garcke, and B. Stinner. Multicomponent alloy solidification: Phase-field modeling and simulations. Physical Review E, 71:1–6, 2005.

[81] B. Nestler, M. Reichardt, and M. Selzer. Massive multi-phase-field simulations: Methods to compute large grain system. In J Hirsch, B Skrotzki, and G Gottstein, editors, Proceedings of the 11th International Conference on Aluminium Alloys, 22 - 26 Sept. 2008, Aachen, Germany, pages 1251–1255. WILEY-VCH Verlag GmbH and Co.KGaA, Weinheim, 2008.

[82] A. Nguyen, N. Satish, and J. Chhugani. 3.5-D blocking optimization for stencil computations on modern CPUs and GPUs. Proceedings of the 2010 ..., pages 1–13, 2010.

[83] Y.J. Park, N.M. Hwang, and D.Y. Yoon. Abnormal growth of faceted (WC) grains in a (Co) liquid matrix. Metallurgical and Materials Transactions, 27, 1996.

[84] D. Parkinson. Parallel efficiency can be greater than unity. Parallel Computing, 3:261–262, 1986.

[85] R.G. Parr. Density Functional Theory. Annual Review of Physical Chemistry, 34:631–656, 1983.

[86] M. Plapp and A. Karma. Multiscale random-walk algorithm for simulating interfacial pattern formation. Physical review letters, 84:1740–1743, 2000.

[87] R.B. Potts. Some generalized order-disorder transformations. Mathematical Proceedings of the Cambridge Philosophical Society, 48:106–109, 1952.

[88] N. Provatas, N. Goldenfeld, and J. Dantzig. Adaptive mesh refinement computation of solidification microstructures using dynamic data structures. Journal of Computational Physics, 148:265–290, 1999.

[89] D. Raabe. Introduction of a scalable three-dimensional cellular automaton with a probabilistic switching rule for the discrete mesoscale simulation of recrystallization phenomena. Philosophical Magazine A, 79:2339–2358, 1999.

[90] D. Raabe and R.C. Becker. Coupling of a crystal plasticity finite-element model with a probabilistic cellular automaton for simulating

primary static recrystallization in aluminium. October, 8:445–462, 2000.

[91] D. Raabe and K. Lücke. Rolling and Annealing Textures of BCC Metals. Materials Science Forum, 157-162:597–610, 1994.

[92] W.T. Read. Dislocations in Crystals. McGraw Hill Book Co., 1953.

[93] W.T. Read and W. Shockley. Dislocation Models of Crystal Grain Boundaries. Physical Review, 78:275–289, 1950.

[94] M. Reichardt, M. Selzer, B. Nestler, A. Unser, D. Raabe, and A. Khorashadizadeh. Massive multi-phase-field simulations: An efficient algorithm to compute large grain systems with constructed anisotrophy in 3d. The publication has not been submitted during writing this thesis, such that the title of the manuscript may change untill submission., 01 2014.

[95] P.R. Rios, F. Siciliano, H.R.Z. Sandim, and R.L. Plaut. Nucleation and growth during recrystallization. Materials, 8:225–238, 2005.

[96] G.S. Rohrer. Grain boundary energy anisotropy: a review. Journal of Materials Science, 46:5881–5895, 2011.

[97] A.D. Rollett, A.P. Brahme, and C.G. Roberts. An Overview of Accomplishments and Challenges in Recrystallization and Grain Growth. Materials Science Forum, 558-559:33–42, 2007.

[98] A.D. Rollett and W.W. Mullins. On the growth of abnormal grains. Scripta materialia, 36:975–980, 1997.

[99] A.D. Rollett, D.J. Srolovitz, and M.P. Anderson. Simulation and theory of abnormal grain growth-anisotropic grain boundary energies and mobilittes. Acta metallurgica, 37:1227–1240, 1989.

[100] E. Schmid and W. Boas. Kristallplastizität mit besonderer Berücksichtigung der Metalle. Angewandte Chemie, 48, 1935.

[101] K. Schulz, D. Dickel, S. Schmitt, S. Sandfeld, D. Weygand, and P. Gumbsch. Analysis of dislocation pile-ups using a dislocation-based continuum theory. Modelling and Simulation in Materials Science and Engineering, 22:025008, 2014.

[102] A.J. Schwartz, M. Kumar, B. L. Adams, and P. Field. Electron backscatter diffraction in materials science. Springer, 2nd edition, 2009.

[103] M. Selzer, M. Jainta, and B. Nestler. A Lattice-Boltzmann model to simulate the growth of dendritic and eutectic microstructures under the influence of fluid flow. physica status solidi (b), 246:1197–1205, 2009.

[104] S.L. Semiatin and H.R. Piehler. Formability of sandwich sheet materials in plane strain compression and rolling. Metallurgical and Materials Transactions A, 10:97–107, 1979.

[105] T. Shimokawabe, T. Aoki, T. Takaki, T. Endo, A. Yamanaka, N. Maruyama, A. Nukada, and S. Matsuoka. Peta-scale phase-field simulation for dendritic solidification on the TSUBAME 2.0 supercomputer. In Proceedings of 2011 International Conference for High Performance Computing, Networking, Storage and Analysis, page 3. ACM, 2011.

[106] E.J. Siem, W.C. Carter, and D. Chatain. The equilibrium shape of anisotropic interfacial particles. Philosophical Magazine, 84:991–1010, 2004.

[107] M. Sommer, W.D. Schubert, E. Zobetz, and P. Warbichler. On the formation of very large WC crystals during sintering of ultrafine WC-Co alloys. International Journal of Refractory Metals and Hard Materials, 20:41–50, 2002.

[108] R. Spatschek, C. Müller-Gugenberger, E. Brener, and B. Nestler. Phase field modeling of fracture and stress-induced phase transitions. Physical Review E, 75:1–14, 2007.

[109] D.J. Srolovitz, G.S. Crest, and M.P. Anderson. Computer simulation of recrystallization–I. Homogeneous nucleation and growth. Acta Metallurgica et Materialia, 34:1833–1845, 1986.

[110] Y. Suwa, Y. Saito, and H. Onodera. Phase-field simulation of recrystallization based on the unified subgrain growth theory. Computational Materials Science, 44:286–295, 2008.

[111] T. Takaki, Y. Hisakuni, T. Hirouchi, A. Yamanaka, and Y. Tomita. Multi-phase-field simulations for dynamic recrystallization. Computational Materials Science, 45:881–888, 2009.

[112] T. Takaki and Y. Tomita. Static recrystallization simulations starting from predicted deformation microstructure by coupling multi-phase-field method and finite element method based on crystal plasticity. International Journal of Mechanical Sciences, 52:320–328, 2010.

[113] G.I. Taylor. Plastic strain in metals. Journal Of The Institute Of Metals, 62:307–324, 1938.

[114] J.E. Taylor and J.W. Cahn. Diffuse interfaces with sharp corners and facets: Phase field models with strongly anisotropic surfaces. Physica D: Nonlinear Phenomena, 112:381–411, 1998.

[115] T. Uehara and R.F. Sekerka. Phase field simulations of faceted growth for strong anisotropy of kinetic coefficient. J. Cryst. Growth, 254:251–261, 2003.

[116] S. Vedantam and B.S.V. Patnaik. Efficient numerical algorithm for multiphase field simulations. Physical Review E - Statistical, Nonlinear and Soft Matter Physics, 73:016703, 2006.

[117] L. Vitos, A.V. Ruban, H.L. Skriver, and J. Kollar. The surface energy of metals. Surface Science, 411:186–202, 1998.

[118] J. Von Neumann. No Title. Metal Interfaces (ed. Herring, C.), pages 108–110, 1952.

[119] John von Neumann. Theory of Self-Reproducing Automata, 1966.

[120] A. Vondrous, M. Reichardt, and B. Nestler. Growth rate distributions for regular two-dimensional grains with read–shockley grain boundary energy. Modelling and Simulation in Materials Science and Engineering, 22, 2014.

[121] A. Vondrous, M. Selzer, J. Hotzer, and B. Nestler. Parallel computing for phase-field models. International Journal of High Performance Computing Applications, 2013.

[122] J.A. Warren and W.J. Boettinger. Prediction of dendritic growth and microsegregation patterns in a binary alloy using the phase-field method. Acta Metallurgica et Materialia, 43:689–703, 1995.

[123] R. Wawszczak, A. Baczmański, C. Braham, W. Seiler, M. Wróbel, and K. Wierzbanowski. Evolution of Residual Stresses and Stored Elastic Energy in Ferritic Steel during Recovery Process. Materials Science Forum, 652:279–284, 2010.

[124] C. Wegst and M. Wegst. Stahlschlüssel - Key to Steel 2010. Stahlschlüssel Wegst GmbH, 2010.

[125] F. Wendler, C. Mennerich, and B. Nestler. A phase-field model for polycrystalline thin film growth. Journal of Crystal Growth, 327:189–201, 2011.

[126] D. Weygand, Y. Bréchet, and J. Lépinoux. A vertex dynamics simulation of grain growth in two dimensions. Philosophical Magazine Part B, 78:329–352, 1998.

[127] D. Weygand, Y. Bréchet, J. Lépinoux, and W. Gust. Three-dimensional grain growth: A vertex dynamics simulation. Philosophical Magazine Part B, 79:703–716, 1999.

[128] H. Won Lee and Y.T. Im. Numerical modeling of dynamic recrystallization during nonisothermal hot compression by cellular automata and finite element analysis. International Journal of Mechanical Sciences, 52:1277–1289, 2010.

[129] G. Wulff. Zur frage der geschwindigkeit des wachstums und der auflösung der krystallflächen. Zeitschrift für Krystallographie und Mineralogie, 34:449–530, 1901.

[130] N. Yamaki, Y. Aoyagi, and K. Shizawa. Multiscale Modeling and Simulation of Crystal Plasticity Based on Dislocation Patterning in Polycrystal. Key Engineering Materials, 340-341:205–210, 2007.

[131] B. Yoon, B. Lee, and S. Kang. Growth behavior of rounded (Ti,W)C and faceted WC grains in a Co matrix during liquid phase sintering. Acta Materialia, 53:4677–4685, 2005.

[132] T. Young. An essay on the cohesion of fluids. Philosophical Transactions of the Royal Society of London, 95:65–87, 1805.

[133] A. Yuichiro, T. Yuzo, I. Tomohiro, H. Shinya, and S. Toshiyuki. The Tofu Interconnect. In 2011 IEEE 19th Annual Symposium on High Performance Interconnects, pages 87–94. IEEE, 2011.

Schriftenreihe des Instituts für Angewandte Materialien

ISSN 2192-9963

Die Bände sind unter www.ksp.kit.edu als PDF frei verfügbar oder als Druckausgabe bestellbar.

Band 1 Prachai Norajitra
Divertor Development for a Future Fusion Power Plant. 2011
ISBN 978-3-86644-738-7

Band 2 Jürgen Prokop
Entwicklung von Spritzgießsonderverfahren zur Herstellung von Mikrobauteilen durch galvanische Replikation. 2011
ISBN 978-3-86644-755-4

Band 3 Theo Fett
New contributions to R-curves and bridging stresses – Applications of weight functions. 2012
ISBN 978-3-86644-836-0

Band 4 Jérôme Acker
Einfluss des Alkali/Niob-Verhältnisses und der Kupferdotierung auf das Sinterverhalten, die Strukturbildung und die Mikrostruktur von bleifreier Piezokeramik $(K_{0,5}Na_{0,5})NbO_3$. 2012
ISBN 978-3-86644-867-4

Band 5 Holger Schwaab
Nichtlineare Modellierung von Ferroelektrika unter Berücksichtigung der elektrischen Leitfähigkeit. 2012
ISBN 978-3-86644-869-8

Band 6 Christian Dethloff
Modeling of Helium Bubble Nucleation and Growth in Neutron Irradiated RAFM Steels. 2012
ISBN 978-3-86644-901-5

Band 7 Jens Reiser
Duktilisierung von Wolfram. Synthese, Analyse und Charakterisierung von Wolframlaminaten aus Wolframfolie. 2012
ISBN 978-3-86644-902-2

Band 8 Andreas Sedlmayr
Experimental Investigations of Deformation Pathways in Nanowires. 2012
ISBN 978-3-86644-905-3

Band 9 Matthias Friedrich Funk
Microstructural stability of nanostructured fcc metals during cyclic deformation and fatigue. 2012
ISBN 978-3-86644-918-3

Band 10 Maximilian Schwenk
Entwicklung und Validierung eines numerischen Simulationsmodells zur Beschreibung der induktiven Ein- und Zweifrequenzrandschicht-härtung am Beispiel von vergütetem 42CrMo4. 2012
ISBN 978-3-86644-929-9

Band 11 Matthias Merzkirch
Verformungs- und Schädigungsverhalten der verbundstranggepressten, federstahldrahtverstärkten Aluminiumlegierung EN AW-6082. 2012
ISBN 978-3-86644-933-6

Band 12 Thilo Hammers
Wärmebehandlung und Recken von verbundstranggepressten Luftfahrtprofilen. 2013
ISBN 978-3-86644-947-3

Band 13 Jochen Lohmiller
Investigation of deformation mechanisms in nanocrystalline metals and alloys by in situ synchrotron X-ray diffraction. 2013
ISBN 978-3-86644-962-6

Band 14 Simone Schreijäg
Microstructure and Mechanical Behavior of Deep Drawing DC04 Steel at Different Length Scales. 2013
ISBN 978-3-86644-967-1

Band 15 Zhiming Chen
Modelling the plastic deformation of iron. 2013
ISBN 978-3-86644-968-8

Band 16 Abdullah Fatih Çetinel
Oberflächendefektausheilung und Festigkeitssteigerung von niederdruckspritzgegossenen Mikrobiegebalken aus Zirkoniumdioxid. 2013
ISBN 978-3-86644-976-3

Band 17 Thomas Weber
Entwicklung und Optimierung von gradierten Wolfram/EUROFER97-Verbindungen für Divertorkomponenten. 2013
ISBN 978-3-86644-993-0

Band 18 Melanie Senn
Optimale Prozessführung mit merkmalsbasierter Zustandsverfolgung. 2013
ISBN 978-3-7315-0004-9

Band 19 Christian Mennerich
Phase-field modeling of multi-domain evolution in ferromagnetic shape memory alloys and of polycrystalline thin film growth. 2013
ISBN 978-3-7315-0009-4

Band 20 Spyridon Korres
On-Line Topographic Measurements of Lubricated Metallic Sliding Surfaces. 2013
ISBN 978-3-7315-0017-9

Band 21 Abhik Narayan Choudhury
Quantitative phase-field model for phase transformations in multi-component alloys. 2013
ISBN 978-3-7315-0020-9

Band 22 Oliver Ulrich
Isothermes und thermisch-mechanisches Ermüdungsverhalten von Verbundwerkstoffen mit Durchdringungsgefüge (Preform-MMCs). 2013
ISBN 978-3-7315-0024-7

Band 23 Sofie Burger
High Cycle Fatigue of Al and Cu Thin Films by a Novel High-Throughput Method. 2013
ISBN 978-3-7315-0025-4

Band 24 Michael Teutsch
Entwicklung von elektrochemisch abgeschiedenem LIGA-Ni-Al für Hochtemperatur-MEMS-Anwendungen. 2013
ISBN 978-3-7315-0026-1

Band 25 Wolfgang Rheinheimer
Zur Grenzflächenanisotropie von $SrTiO_3$. 2013
ISBN 978-3-7315-0027-8

Band 26 Ying Chen
Deformation Behavior of Thin Metallic Wires under Tensile and Torsional Loadings. 2013
ISBN 978-3-7315-0049-0

Band 27 Sascha Haller
Gestaltfindung: Untersuchungen zur Kraftkegelmethode. 2013
ISBN 978-3-7315-0050-6

Band 28 Stefan Dietrich
Mechanisches Verhalten von GFK-PUR-Sandwichstrukturen unter quasistatischer und dynamischer Beanspruchung. 2013
ISBN 978-3-7315-0074-2

Band 29 Gunnar Picht
Einfluss der Korngröße auf ferroelektrische Eigenschaften dotierter $Pb(Zr_{1-x}Ti_x)O_3$ Materialien. 2013
ISBN 978-3-7315-0106-0

Band 30 Esther Held
Eigenspannungsanalyse an Schichtverbunden mittels inkrementeller Bohrlochmethode. 2013
ISBN 978-3-7315-0127-5

Band 31 Pei He
On the structure-property correlation and the evolution of Nanofeatures in 12-13.5% Cr oxide dispersion strengthened ferritic steels. 2014
ISBN 978-3-7315-0141-1

Band 32 Jan Hoffmann
Ferritische ODS-Stähle – Herstellung, Umformung und Strukturanalyse. 2014
ISBN 978-3-7315-0157-2

Band 33 Wiebke Sittel
Entwicklung und Optimierung des Diffusionsschweißens von ODS Legierungen. 2014
ISBN 978-3-7315-0182-4

Band 34 Osama Khalil
Isothermes Kurzzeitermüdungsverhalten der hoch-warmfesten Aluminium-Knetlegierung 2618A (AlCu2Mg1,5Ni). 2014
ISBN 978-3-7315-0208-1

Band 35 Magalie Huttin
Phase-field modeling of the influence of mechanical stresses on charging and discharging processes in lithium ion batteries. 2014
ISBN 978-3-7315-0213-5

Band 36 Christoph Hage
Grundlegende Aspekte des 2K-Metallpulverspritzgießens. 2014
ISBN 978-3-7315-0217-3

Band 37 Bartłomiej Albiński
Instrumentierte Eindringprüfung bei Hochtemperatur für die Charakterisierung bestrahlter Materialien. 2014
ISBN 978-3-7315-0221-0

Band 38 Tim Feser
Untersuchungen zum Einlaufverhalten binärer alpha-Messinglegierungen unter Ölschmierung in Abhängigkeit des Zinkgehaltes. 2014
ISBN 978-3-7315-0224-1

Band 39	Jörg Ettrich **Fluid Flow and Heat Transfer in Cellular Solids.** 2014 ISBN 978-3-7315-0241-8
Band 40	Melanie Syha **Microstructure evolution in strontium titanate Investigated by means of grain growth simulations and x-ray diffraction contrast tomography experiments.** 2014 ISBN 978-3-7315-0242-5
Band 41	Thomas Haas **Mechanische Zuverlässigkeit von gedruckten und gasförmig abgeschiedenen Schichten auf flexiblem Substrat.** 2014 ISBN 978-3-7315-0250-0
Band 42	Aron Kneer **Numerische Untersuchung des Wärmeübertragungsverhaltens in unterschiedlichen porösen Medien.** 2014 ISBN 978-3-7315-0252-4
Band 43	Manuel Feuchter **Investigations on Joule heating applications by multiphysical continuum simulations in nanoscale systems.** 2014 ISBN 978-3-7315-0261-6
Band 44	Alexander Vondrous **Grain growth behavior and efficient large scale simulations of recrystallization with the phase-field method.** 2014 ISBN 978-3-7315-0280-7